METHODS IN MOLECULAR BIOLOGY™

Series Editor
John M. Walker
School of Life Sciences
University of Hertfordshire
Hatfield, Hertfordshire, AL10 9AB, UK

For further volumes:
http://www.springer.com/series/7651

Planar Cell Polarity

Methods and Protocols

Edited by

Kursad Turksen

*Regenerative Medicine Program, Sprott Centre for Stem Cell Research,
Ottawa Hospital Research Institute, Ottawa, ON, Canada*

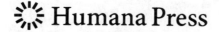 Humana Press

Editor
Kursad Turksen, Ph.D.
Regenerative Medicine Program
Sprott Centre for Stem Cell Research
Ottawa Hospital Research Institute
Ottawa, ON, Canada
kturksen@ohri.ca

Additional material to this book can be downloaded from http://extras.springer.com

ISSN 1064-3745   e-ISSN 1940-6029
ISBN 978-1-61779-509-1   e-ISBN 978-1-61779-510-7
DOI 10.1007/978-1-61779-510-7
Springer New York Dordrecht Heidelberg London

Library of Congress Control Number: 2011943888

Printed on acid-free paper

Humana Press is part of Springer Science+Business Media (www.springer.com)

Preface

With the ever-increasing awareness of the importance of *Planar Cell Polarity* in not only normal development but also disease states, I felt that it was timely to gather a number of protocols related to this topic.

I believe that this collection of protocols will be valuable to both novices and experts in the field. I thank all of the contributors for their efforts in putting together the details of their favorite protocols. This volume would not have materialized without their efforts. I would also like especially to acknowledge Dr. Peng Chen who was extremely supportive of me at the initiation of this project.

As always, I am grateful to Dr. John Walker for his continuous support and for his commitment to maintaining such a high-quality series. My thanks also to Patrick Marton for his encouragements and help. Finally, a special thank you goes to David Casey for his tremendous help in the production of the volume.

Ottawa, ON, Canada   *Kursad Turksen, Ph.D.*

Contents

Contributors

YOHANNS BELLAICHE • *Polarity Division and Morphogenesis, Institut Curie, CNRS UMR 3215, INSERM U934, Paris, France*

SIMON COLLIER • *Department of Biological Sciences, Marshall University, Huntington, WV, USA*

ANDREW J. COPP • *Neural Development Unit, Institute of Child Health, University College London, London, UK*

IRA O. DAAR • *Laboratory of Cell and Developmental Signaling, National Cancer Institute-Frederick, Frederick, MD, USA*

DARYL J.V. DAVID • *Department of Cell & Systems Biology, University of Toronto, Toronto, ON, Canada*

MITSUHARU ENDO • *Department of Physiology and Cell Biology, Graduate School of Medicine, Kobe University, Kobe, Japan*

ANDREW FORGE • *Centre for Auditory Research, UCL Ear Institute, University College London, London, UK*

LISA FUJIMURA • *Department of Biomedical Science, Graduate School of Medicine, Chiba University, Chiba City, Chiba, Japan*

LAURA GRABEL • *Department of Biology, Wesleyan University, Middletown, CT, USA*

NICHOLAS D.E. GREENE • *Neural Development Unit, Institute of Child Health, University College London, London, UK*

RAYMOND HABAS • *Department of Biochemistry, UMDNJ-Robert Wood Johnson Medical School, Piscataway, NJ, USA; Department of Biology, College of Science and Technology, Temple University, Philadelphia, PA, USA*

JIN-KWAN HAN • *Division of Molecular and Life Sciences, Department of Life Science, Pohang University of Science and Technology, Hyoja Dong, Pohang, Kyungbuk, Republic of Korea*

HESTER HAPPÉ • *Department of Human Genetics and Pathology, Leiden University Medical Center, Leiden, The Netherlands*

TONY J.C. HARRIS • *Department of Cell & Systems Biology, University of Toronto, Toronto, ON, Canada*

MASAHIKO HATANO • *Department of Biomedical Science, Graduate School of Medicine, Chiba University, Chiba City, Chiba, Japan*

EMILE DE HEER • *Department of Pathology, Leiden University Medical Center, Leiden, The Netherlands*

JUSTIN HOGAN • *Department of Biological Sciences, Marshall University, Huntington, WV, USA*

DANIEL J. JAGGER • *Centre for Auditory Research, UCL Ear Institute, University College London, London, UK*

BERTRAND JAUFFRED • *Polarity Division and Morphogenesis, Institut Curie, CNRS UMR 3215, INSERM U934, Paris, France*

JASON R. JESSEN • *Division of Genetic Medicine, Department of Medicine, Vanderbilt University Medical Center, Nashville, TN, USA*

MATTHEW W. KELLEY • *Laboratory of Cochlear Development, National Institute on Deafness and other Communication Disorders, National Institutes of Health, Bethesda, MD, USA*

AKIRA KIKUCHI • *Department of Molecular Biology & Biochemistry, Graduate School of Medicine, Faculty of Medicine, Osaka University, Osaka, Japan*

GUN-HWA KIM • *Division of Life Science, Korea Basic Science Institute, Daejeon, Republic of Korea*

DAVID KIMELMAN • *Department of Biochemistry, University of Washington, Seattle, WA, USA*

YUKO KOMIYA • *Department of Biology, College of Science and Technology, Temple University, Philadelphia, PA, USA*

KRISTI LAMONICA • *Department of Craniofacial Biology, School of Dental Medicine, Anschutz Medical Campus, University of Colorado Denver, Aurora, CO, USA*

HYUN-SHIK LEE • *School of Life Sciences, College of Natural Sciences, Kyungpook National University, Daegu, South Korea*

CHRISTOPHE MARCELLE • *EMBL Australia, Australian Regenerative Medicine Institute (ARMI), Monash University, Clayton, VIC, Australia*

VALENTINA MASSA • *Dulbecco Telethon Institute at M.Tettamanti Research Center, Pediatric Department, University of Milano-Bicocca, Monza, Italy*

SHINJI MATSUMOTO • *Department of Molecular Biology & Biochemistry, Graduate School of Medicine, Faculty of Medicine, Osaka University, Osaka, Japan*

MAKOTO MATSUYAMA • *Division of Biochemistry, Aichi Cancer Center Research Institute, Nagoya, Japan*

HELEN MAY-SIMERA • *Laboratory of Cochlear Development, National Institute on Deafness and other Communication Disorders, National Institutes of Health, Bethesda, MD, USA*

JOHN W. MCAVOY • *Sydney Hospital and Eye Hospital, Sydney, NSW, Australia*

MELANIE A. MCGILL • *Department of Cell & Systems Biology, University of Toronto, Toronto, ON, Canada*

R.F. ANDREW MCKINLEY • *Department of Cell & Systems Biology, University of Toronto, Toronto, ON, Canada*

COURTNEY MEZZACAPPA • *Department of Biochemistry, UMDNJ-Robert Wood Johnson Medical School, Piscataway, NJ, USA*

YASUHIRO MINAMI • *Department of Physiology and Cell Biology, School of Medicine, Kobe University, Kobe, Japan*

SALLY A. MOODY • *Department of Anatomy and Regenerative Biology, The George Washington University Medical Center, Washington, DC, USA*

DAVID NEFF • *Department of Biological Sciences, Marshall University, Huntington, WV, USA*

MICHIRU NISHITA • *Department of Physiology and Cell Biology, Graduate School of Medicine, Kobe University, Kobe, Japan*

EDMOND CHANGKYUN PARK • *Division of Life Science, Korea Basic Science Institute, Daejeon, Republic of Korea*

DORIEN J.M. PETERS • *Department of Human Genetics, Leiden University Medical Center, Leiden, The Netherlands*

SOPHIE E. PRYOR • *Neural Development Unit, Institute of Child Health, University College London, London, UK*

ANNE C. RIOS • *EMBL Australia, Australian Regenerative Medicine Institute (ARMI), Monash University, Clayton, VIC, Australia*

DAWN SAVERY • *Neural Development Unit, Institute of Child Health, University College London, London, UK*

OLIVIER SERRALBO • *EMBL Australia, Australian Regenerative Medicine Institute (ARMI), Monash University, Clayton, VIC, Australia*

AKIHIKO SHIMONO • *Centre of Life Sciences, Cancer Science Institute of Singapore, National University of Singapore, Singapore*

SERGEI Y. SOKOL • *Department of Developmental and Regenerative Biology, Mount Sinai School of Medicine, New York, NY, USA*

YUKI SUGIYAMA • *Save Sight Institute, The University of Sydney, Sydney, NSW, Australia*

YINGQUN WANG • *Department of Pathology, Anatomy, and Cell Biology, Thomas Jefferson University, Philadelphia, PA, USA*

DOUGLAS C. WEISER • *Department of Biological Sciences, University of the Pacific, Stockton, CA, USA*

Chapter 1

Live Imaging of *Drosophila* Embryos: Quantifying Protein Numbers and Dynamics at Subcellular Locations

Daryl J.V. David*, Melanie A. McGill*, R.F. Andrew McKinley*, and Tony J.C. Harris

Abstract

Live imaging is critical for understanding the structure and activities of protein interaction networks in cells. By tagging proteins of interest with fluorescent proteins, such as green fluorescent protein (GFP), their localization in cells can be determined and correlated with cellular activities. This can be extended into developmental systems such as *Drosophila* to understand the molecular and cellular bases of development. In this chapter, we review sample preparation techniques and basic imaging considerations for *Drosophila* embryos. We then discuss how these techniques can be extended to count absolute protein numbers at specific subcellular locations, and determine their dynamics using fluorescence recovery after photobleaching (FRAP). These techniques can help reveal the structure and dynamics of protein complexes in live cells.

Key words: *Drosophila*, Live imaging, Protein counting, Fluorescence recovery after photobleaching, Adherens junctions, Epithelia

1. Introduction

The *Drosophila* embryo is an excellent system for studying the molecular and cellular bases of development through live imaging. In our lab, we are particularly interested in epithelial morphogenesis. We have applied live imaging techniques to study how adherens junctions first assemble between cells as the embryonic epithelium forms at cellularization (1), in addition to epithelial remodeling at later stages (2, 3). In this chapter, we review sample preparation techniques and basic imaging considerations from our lab. We discuss how these techniques can be extended to count

---

*The first three authors contributed equally to this Chapter.

Kursad Turksen (ed.), *Planar Cell Polarity: Methods and Protocols*, Methods in Molecular Biology, vol. 839,
DOI 10.1007/978-1-61779-510-7_1, © Springer Science+Business Media, LLC 2012

protein numbers at specific subcellular locations to reveal the structure of precursory adherens junctions within live cells. Then, we discuss how fluorescence recovery after photobleaching (FRAP) can be used to determine protein dynamics at these complexes. Finally, we comment on statistical considerations when studying cell biology in *Drosophila* embryos.

2. Materials

2.1. Dechorionation of Drosophila Embryos

1. Apple-juice agar plates for collecting embryos.
2. 0.1% Triton X-100 (TX-100) in water.
3. 50% household bleach in water. Diluted bleach solutions break down. To maintain its effectiveness, make this solution in relatively small batches (<250 ml) so that it can be used in 1–2 months.

2.2. Mounting Embryos on Gas-Permeable Petri-PERM Membranes

1. Halocarbon oil series 700 (Halocarbon Products Corporation).
2. Gas-permeable petri-PERM membranes (Greiner Lumox culture dish, 50 mm, hydropholic from Sigma-Aldrich). These gas-permeable membranes, when cleaned promptly with a Kim-wipe after use, can be reused multiple times (when the membrane becomes loose, the sample will move in and out of focus while imaging).
3. Glass coverslips, square (VWR Micro Cover Glasses, 22 mm Square, No. 1½, 0.16–0.19 mm thickness).

2.3. Mounting and Gluing Embryos on Coverslips

1. Embryo glue solution: In a scintillation vial, rock ~2 ml of heptane to dissolve the adhesive from 1 cm of clear adhesive tape overnight at room temperature. Collect supernatant and remove residual debris through centrifugation. This solution can be further diluted in heptane as needed.
2. Dissecting microscope.
3. Open air coverslip holder: a 50 mm petri dish with a ~15 mm diameter hole. The hole can be made by placing a petri dish briefly over a Bunsen burner which melts the centre away. The dish is then quickly flattened with pressure from a capped 50 ml Falcon tube.

2.4. Counting GFP Protein Numbers Using ELISAs

1. Purified GST-GFP.
2. BCA protein assay (Thermo Fisher Scientific).
3. Bovine serum albumin (BSA).
4. Mini glass homogenizer.

5. NP-40 lysis buffer (150 mM sodium chloride, 1 μg/ml aprotinin, 1 μg/ml leupeptin, 1 μg/ml pepstatin, 1 μg/ml PMSF, 1.0% NP-40, and 50 mM Tris–HCl, pH 8.0).

6. ELISA plates coated with goat anti-GFP antibodies (Thermo Fisher Scientific).

7. Rabbit anti-GFP antibodies (ab290; Abcam).

8. Goat anti-rabbit-HRP antibodies (Thermo Fisher Scientific).

9. Detection reagent (1-Step Ultra TMB-ELISA; Thermo Fisher Scientific).

10. ELISA plate reader.

3. Methods

3.1. Sample Preparation and Basic Imaging Considerations

Tagging proteins with fluorophores, such as GFP, can reveal their dynamics and subcellular localization in a living organism or cell. The following section outlines steps for sample preparation and basic live imaging in *Drosophila* embryos (see Note 1).

3.1.1. Dechorionation of Drosophila Embryos

1. With a paintbrush pre-moistened with 0.1% TX-100, collect *Drosophila* embryos from apple juice–agar collection plates by sweeping them into groups and transferring them into a 1.5 ml tube of 0.1% TX-100.

2. After embryos settle to the bottom of the tube, wash embryos 2–3 times in 0.1% TX-100, inverting each time and removing all supernatant and liquid from the caps of tubes. This should be done until the yeast is adequately removed such that the 0.1% TX-100 in the tube is clear.

3. Remove the chorion by rocking in 50% bleach for 3 min, then allowing the embryos to settle for 1 min. In this way, embryos are not exposed to the bleach solution for more than 4 min total.

4. Wash embryos three times in 0.1% TX-100, inverting each time and removing all supernatant and liquid from the caps of tubes (see Note 2).

3.1.2. Mounting Embryos on a Gas-Permeable Petri-PERM Membrane

This method is relatively quick and easy. Also, a slight compression of the embryos (imposed by the membrane and coverslip) flattens the sample slightly to allow a greater area of the embryo surface to be imaged within the same focal plane. However, the increased pressure can affect specimens adversely with prolonged imaging.

1. Enlarge the hole of a p200 tip by cutting its end with scissors. Aspirate 200 μl of dechorionated embryos in 0.1% TX-100 into the tip.

2. Place the wet embryos onto a petri-PERM dish in a relatively even and disperse fashion. Embryos that are too densely packed can pile on top of one another. This also increases the difficulty in adequately removing the 0.1% TX-100 from each embryo.

3. Excess pools of liquid can be first removed with a p200 micropipettor. Then, dry the embryos with twisted corners of Kim-Wipes. Touch each embryo to remove halos of 0.1% TX-100 surrounding the embryos (Fig. 1a), and then air dry for 3–5 min before applying halocarbon oil. This is a critical step as excess 0.1% TX-100 obscures fluorescent signals while imaging.

4. Use a glass rod to place a small drop of halocarbon oil on a coverslip and place on top of the dried embryos (Fig. 1b). Excess halocarbon oil will decrease the compression necessary

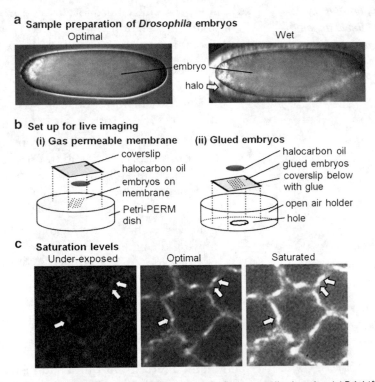

Fig. 1. Preparation of *Drosophila* embryo samples for fluorescent live imaging. (**a**) Brightfield images of live embryos mounted in Halocarbon oil with sufficient and insufficient removal of 0.1% TX-100 (*optimal* and *wet*, respectively). Images taken with a ×20 objective. Note the halo of liquid around the wet embryo. (**b**) Schematic set up of *Drosophila* embryos for live imaging either (i) mounted on a gas-permeable membrane or (ii) glued to a coverslip. (**c**) Spinning disc confocal images of DE-cad::GFP in a cellularizing *Drosophila* embryo with a ×63 objective at different exposure times. The embryos are underexposed, optimally exposed, and saturated. Note that the DE-cad::GFP puncta (*arrows*) are difficult to resolve when saturated.

for holding embryos in place and flattening them. The drop of halocarbon oil should cover ~33% of the coverslip before being applied to the embryos. Once placed on the embryos, the oil takes about a minute to spread under the coverslip. It is ideal if the oil does not fully spread out underneath the entirety of the coverslip.

3.1.3. Mounting and Gluing Embryos on a Coverslip

Although gluing individual embryos onto a coverslip can be more labour-intensive than placing them on a gas-permeable membrane, gluing embryos directly on coverslips poses several advantages for long-term imaging. First, as embryos are not sandwiched between both a coverslip and a membrane, they are not subject to compression which may adversely affect their behaviour. Second, since embryos are covered in halocarbon oil and exposed to air, there is a decreased risk of developing anoxic conditions. Also, embryos can be pre-selected and oriented on an agar block prior to gluing on a coverslip (embryos can also be manually oriented on petri-PERM dishes).

1. Allow apple juice–agar plates to sit at room temperature for 2–3 days to sufficiently dry them for embryonic manipulations.

2. Cut several blocks of apple juice–agar from the plate. A block of ~1 cm × 3 cm provides sufficient space to allow for embryonic manipulations.

3. Enlarge the hole of a p200 tip by cutting its end with scissors. Aspirate 200 µl of dechorionated embryos and 0.1% TX-100 into the tip.

4. Pile dechorionated embryos onto pre-cut agar blocks and remove excess TX-100.

5. Using a dissecting needle under a dissecting microscope, pick up groups of desired embryos and place them onto a new agar block, nudging them gently to align and rotate them as desired. Excessive force may rupture embryos. Rotate embryos so the side to be imaged is facing up from the agar.

6. Dip a p200 tip into a tube containing embryo glue solution and apply a thin layer of embryo glue to a coverslip. Allow for the heptane to evaporate (less than 1 min). As thin a layer as possible should be applied, since the glue has some auto-fluorescence that can obscure imaging. An additional layer of embryo glue can be applied if there is insufficient adhesiveness (test by touching the corner of the coverslip with a gloved fingertip or other object).

7. Gently press the coverslip, glue side down, onto the agar block with aligned embryos. Gentle contact is sufficient to stick the embryos to the coverslip.

8. Place the coverslip, embryo side up, onto an open air coverslip holder (see Fig. 1b for a schematic). A thin application of halocarbon oil to the rim of the coverslip holder hole helps hold the coverslip in place.

9. Cover embryos with halocarbon oil.

3.1.4. Basic Imaging Considerations

Optimal images contain a wide range of signal intensities without saturation (see Note 3). This can be achieved by balancing the following parameters at the microscope.

(a) *Laser power*: Laser power increases the intensity of the excitation light and thus the intensity of light emitted by the sample. However, increasing laser power also increases photobleaching and phototoxicity. Generally, laser power is kept as low as possible to avoid these effects.

(b) *Exposure time*: Increasing exposure time increases the number of photons contributing to each pixel in the final image. However, increased exposure times can also increase photobleaching and phototoxicity since it increases exposure to excitation lasers.

(c) *Detector sensitivity*: Another method to increase fluorescent signal is to increase the sensitivity of the detector. This avoids photobleaching and phototoxicity, but increased detector sensitivity increases the amount of noise collected in an image.

(d) *z-Resolution and time resolution*: Reconstructing and more carefully analyzing subcellular structures often requires three-dimensional time lapse microscopy. Confocal microscopes allow z-sectioning, and one can control for the number of focal planes (z-sections) and distance between each z-section. Too many z-sections reduce temporal resolution, while too large a distance between z-slices may bypass certain subcellular structures. A 0.3 μm step size between z-slices with a ×63 objective is suitable for most experiments.

3.2. Counting Protein Numbers at Subcellular Locations

To understand how any molecular machinery functions, it is important to know the numbers of its component parts. To determine the numbers of a fluorescently tagged protein at a subcellular location [e.g. spot adherens junctions (SAJs)], two measurements can be used: (1) the total and local relative fluorescence of the protein within the cell and (2) the number of molecules of the protein in the cell (4, 5). When combined, these two measurements indicate the percentage of the total cellular numbers of a protein that resides at a specific site within the cell.

This methodology can be applied to single cells, such as yeast (4), and simple multi-cellular systems such as *Drosophila* embryos at cellularization (1). *Drosophila* cellularization is an ideal multicellular system for this approach because all cells have the same hexagonal shape and dimensions and the total number of cells in the embryo is known (6).

For the counts to reflect the numbers of endogenous proteins at the same site in a wildtype embryo, the sample should ideally contain only the fluorescently tagged version of the protein (in a gene trap or mutant rescue context). Also, it is important to perform a western blot to compare total protein levels between embryos expressing the fluorescently tagged protein of interest and wildtype embryos expressing the endogenous protein.

3.2.1. Making Cortical Localization Maps to Determine the Total and Local Fluorescence of a Protein Within the Cell

1. To prepare embryos for live imaging, collect embryos expressing the fluorescently tagged protein at cellularization as in Subheadings 3.1.1 and 3.1.2.

2. Obtain confocal z-stacks of late cellularization embryos covering the full depth of a cell with a ×63 objective and 0.3 μm step sizes.

3. Export the unaltered z-stack as a TIFF and open in *ImageJ*.

4. It is easiest to make cortical localization maps of single sides of the hexagonal cells. In *ImageJ*, use the *Rectangular Selection Tool* to select one cell side that is relatively straight along the apical–basal axis. Adjust the rectangle size so that the whole membrane is captured from apical to basal. To avoid overestimating the number of SAJs within a cell, only choose one of the two tricellular junctions when cropping the membrane (Fig. 2a, white box). Click *Image>Crop*.

5. To visualize and analyze the distribution of the protein along the entire apical–basal membrane of the single cell side, click *Image>Stacks>Make Montage* to create a montage with one column and the number of rows that are required to cover the entire depth of the cell (a portion of such a montage is shown in Fig. 2b). The following parameters can be determined from the montage; (1) the width and depth of one cell side (knowing the number of microns per pixel in x–y and the step-size in z), (2) the width and depth of SAJs, and (3) the number of SAJs per one cell side and the distribution of SAJs down one cell side.

6. To generate raw intensity traces for individual SAJs, use the straight line tool to trace an SAJ by starting from before, and extending after, where its fluorescence is detectable (Fig. 2b, solid line). Select *Analyze>Set Measurements* and then click the *Mean Grey Value* box. This will measure the greyscale value for each individual pixel along the line. Select *Analyze>Plot Profile* to obtain an intensity distribution versus distance (Fig. 2d). Click *list* in the top left corner, right click on the data, and copy into a spreadsheet. In addition, calculate the length of each SAJ along the z-axis from the number of z-sections it is detected in.

7. Repeat the same measurements for the intervening membrane (IM) using the straight line tool to measure the mean intensity values from apical to basal (Fig. 2b, dotted line). Paste the results into a spreadsheet.

8. To measure the cytoplasm intensity, draw a box in the centre of the cell in the full z-stack and measure its *Mean Grey Value* from apical to basal using the *MeasureStack* plug-in. Paste the results into a spreadsheet, align relative to the top of the cell and average for each embryo.

9. Since the straight line tool passes through both a SAJ and cytoplasm on either side in one z-step (Fig. 2c), selecting the maximum intensity value from that z-step will remove the cytoplasm measurements. To do this for SAJs and IMs, create a new column in the spreadsheet which includes only the maximum intensity values for each z-step [the peak values on the raw intensity traces (Fig. 2d)].

10. Next, plot the intensity profiles for all SAJs in the same embryo using only the peak values from each z-step of the raw traces (Fig. 2d). This converts the raw traces into intensity profiles for each SAJ. Align all of the SAJ profiles so that the maximum value of each sample is within the same row of the spreadsheet (Fig. 2e shows an example of two profiles). If some SAJ profiles are longer than others, lengthen the shorter SAJ profiles by filling in the blank space with the average IM intensity for that part of the cell cortex (Fig. 2e'). These profiles are then averaged (Fig. 2e''). The average IM profile for each embryo is calculated after aligning all of the single IM profiles (Fig. 2f shows an example of two profiles and their average is shown in Fig 2f').

11. To combine data measured from different embryos, normalize the averaged SAJ, IM and cytoplasmic profiles by dividing all of the data points obtained from the same embryo by the highest intensity value of the three averaged profiles [Fig 2e'' (SAJ), Fig 2f'' (IM)].

12. To remove background, subtract the normalized cytoplasmic profiles from the normalized SAJ (Fig. 2e''') and IM (Fig. 2f''') profiles (see Note 4).

13. Calculate the area under the curve for an average SAJ by taking the sum of all of the intensity values within the profile. To determine the total relative cortical fluorescence from all SAJs within the cell, multiply the area under an average SAJ profile by the average SAJ width and then by the average number of SAJs/cell.

14. To determine the total relative cortical fluorescence for IM, calculate the regions of an average cell cortex not occupied by SAJs and apply the mean relative fluorescence intensity profiles of IM

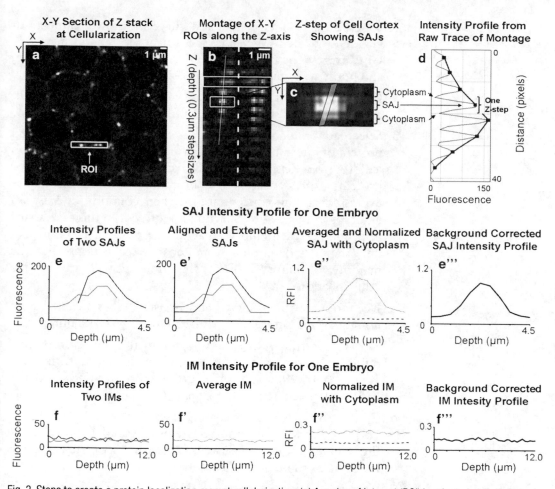

Fig. 2. Steps to create a protein localization map at cellularization. (**a**) A region of interest (ROI) is selected at the cell cortex at cellularization. The white box covers one cell side and one of the two tricellular junctions. (**b**) A montage is created from the ROI at multiple *z* positions. The montage of the ROI shows three SAJs (*white streaks*), with one being selected using the straight line tool. Selection of the intervening membrane is shown with a *dotted line*. (**c**) A single *z*-step from the montage is shown with a line intersecting a SAJ with cytoplasm on either side. (**d**) An example of a raw trace of a SAJ from a montage. The peaks of the graph correspond to where the line intersects the core of the SAJ in each *z*-step (*bracket*). The valleys correspond to the cytoplasm on either side. When connected, the peaks form the fluorescence intensity profile for the SAJ. (**e**, **f**) Raw fluorescence intensity profiles versus depth are plotted for SAJs and IM. The maximum values for all SAJ profiles are aligned. (e′) If some SAJ profiles are longer than others, the missing spaces are filled with the average IM intensity for that part of the cell. (**f′**) The overall average IM distribution for a single embryo is shown. (e″, **f″**) All SAJ, IM, and cytoplasm profiles are normalized by dividing by the highest intensity value of the averaged profiles from the embryo. The data is now in relative fluorescence units. (e‴, **f‴**) To correct for background fluorescence, the normalized cytoplasm profile is subtracted from the normalized SAJ and IM profiles. The background corrected intensity profiles from different embryos can then be combined for SAJs and IM.

to these regions. The total area of the cell cortex is determined by multiplying the mean depth of a cell side by the mean width of a cell side by the six sides of the cell. Combining the total relative cortical fluorescence from SAJs and IM gives an overall map of relative cortical fluorescence of an average cell (1).

1. To create a GST-GFP standard curve of known numbers of GST-GFP proteins, generate a construct for expressing GST-GFP, express the construct in DL21 cells, lyse with lysozyme, and purify with glutathione resin (see Note 5).

2. Quantify the purified GST-GFP protein with a BCA protein assay (Thermo Fisher Scientific) versus a BSA standard curve.

3. Select late cellularization embryos expressing the GFP-tagged protein of interest under a dissecting microscope and place on ice (to stop development). Ten embryos are sufficient for proteins expressed at high concentrations. If the protein is expressed at low concentrations, then counting at least 50 embryos from a 3–4 h collection is required. In this case, visual inspection of each embryo may not be possible and thus it is important to determine from separate collections that the majority of embryos from this time period are at late cellularization.

4. Dechorionate the embryos and transfer to a prechilled mini homogenizer. On ice, lyse with 100 μl NP-40 lysis buffer.

5. At 4°C, centrifuge the lysates for 6 min at $1,050 \times g$, and then transfer the supernatant to a new tube, and centrifuge for 1 min at $16,900 \times g$. Load the full lysate volumes into the ELISA plates coated with goat anti-GFP antibody.

6. Create a GFP standard curve by serial dilution of the pure GST-GFP in wildtype embryo lysates prepared as above. Load the samples into a strip of wells on the ELISA plate alongside the samples to be quantified.

7. Perform the ELISA protocol at 4°C following the supplier's instructions and determine the number of GFP proteins per embryo using the GFP standard curve. Since there are $5,952 \pm 329$ cells per embryo at cellularization (6), the number of GFP-tagged proteins per cell can be derived.

With the total number of molecules for each protein within a cell determined (Subheading 3.2.2) and the relative distribution of the protein(s) within the cell mapped (Subheading 3.2.1), the two pieces of data can be combined to determine the absolute number of molecules at sub-cellular locations within the cell.

1. To determine the number of proteins per SAJ, multiply the counts of proteins per cell by the fraction of total relative cortical fluorescence from SAJs, and then divide by the number of SAJs per cell.

2. Similarly, to determine the number of proteins in the IM, multiply the counts of proteins per cell by the fraction of the total relative cortical fluorescence from IM.

3. Calculate standard deviations for each final mean by dividing the standard deviation of each parameter used by its individual mean, squaring the value, summing these squared values for all parameters used, taking the square root of this sum, and multiplying this value by the final mean.

3.3. Measuring Protein Dynamics Using FRAP

FRAP is a widely used technique to directly study protein dynamics and behaviour *in vivo* (for reviews, see refs. 7, 8). Information can be gained by comparing the kinetics of different molecules or comparing identical molecules under different conditions. In a typical FRAP experiment, a small region of fluorescent molecules in a living cell is permanently bleached and the fluorescence recovery into this region monitored over time. FRAP relies on irreversibility of the bleaching such that molecular movement is reflected in the exchange of fluorescence between bleached and unbleached regions.

FRAP experiments depend on multiple parameters. Therefore, controls and optimal conditions must be worked out for each experiment. Once these conditions have been established, image acquisition and bleaching are routine and easily comparable from experiment to experiment. Here, we discuss conditions and controls for using FRAP to examine protein dynamics in live *Drosophila* embryos. Embryos used in FRAP experiments are prepared for live cell imaging as discussed in Subheadings 3.1.1 and 3.1.2.

3.3.1. Determining Image Acquisition Parameters

(a) *Maximizing time resolution*: As FRAP experiments monitor molecular dynamics, high time-resolution is necessary. Images should be acquired as fast as possible to capture recovery of fluorescent molecules into the bleached region, but long enough to maintain acceptable image quality. A balance between the time resolution and the quality of image must be achieved.

(b) *Minimizing acquisition photobleaching*: The intensity of the acquisition laser is also important and should be adjusted to minimize the amount of undesirable photobleaching resulting from laser exposure during image acquisition. Acquisition photobleaching is corrected for during image analysis as discussed in "Correcting for Intensity Changes Due to Acquisition Photobleaching and Laser Fluctuations" in Subheading 3.3.5.

3.3.2. Selecting the Region to Bleach

(a) *Determining the dimensions*: A number of factors must be considered when selecting the region to bleach. If the selected region is too large, then there may be limited number of molecules available for recovery. If it is too small, then recovery may be too quick to observe. The size and shape of the bleach region in *x–y* is set by drawing a region of interest (ROI) using the imaging software program (Fig. 3a). The dimensions in

z can be examined in fixed embryos where there is no molecular movement—the fixed embryos are bleached using the established conditions and the depth of bleach can be seen in a z-stack (Fig. 3b).

(b) *Adjusting for movement*: During FRAP experiments over time scales longer than a few seconds or in very dynamic cellular processes, cellular movements and movements of the overall embryo can lead to movements of the bleached and unbleached regions being analyzed. Thus, recovery may not be detected because the structure has moved out of the focal plane. To overcome cellular movements, z-stacks can be acquired to accommodate movement along the z plane and the in-focus frame analyzed. The added time required to acquire a z-stack must be balanced with the minimum time resolution needed. Small movements of the ROI in and out of frame for short periods of time may be tolerated in the final image analysis, but manual corrections of sample drift may be necessary [in either case, a neighbouring unbleached structure should be used as a reference (Fig. 3c)]. It may also be necessary to restrict analyses to embryonic stages when cellular movement is minimal.

3.3.3. Establishing Bleach Time

The bleach time is the time an embryo is exposed to the bleaching laser to provide good contrast between the bleached and unbleached regions. Bleach times of 1 s are typical in *Drosophila* embryos which are multiple cells thick and are covered with an outer vitelline membrane. Often bleaching is not 100% complete; however, a 70–80% loss in fluorescence is sufficient to monitor recovery.

The bleach time should be as short as possible to obtain sufficient bleaching while minimizing both the exchange of fluorescent molecules during the bleach time and photodamage. To assess whether molecular exchange occurs during the bleach time, the bleach region in live embryos can be compared to those in fixed embryos. If the edges of the bleached region in the live samples appear fuzzier than the fixed embryos, then exchange will occur and can be corrected by decreasing the bleach time until the live and fixed regions are similar. To assess photodamage, embryos should be monitored for dramatic morphological changes upon bleaching.

3.3.4. Image Acquisition

For subsequent image analysis, sufficient images must be acquired before and after bleaching. The steady-state level of the fluorescent molecules must be established by acquiring adequate pre-bleach images (3–10 images). Similarly, it is important to allow sufficient time for fluorescent molecules to recover ensuring slow recovery is not missed. To determine when to stop imaging, initial experiments should acquire images until there is no noticeable change in fluorescence observed.

3.3.5. Image Analysis

There are five main steps involved in analyzing images from FRAP experiments:

Measuring Fluorescence Intensity Within the Region of Interest

1. Import image of one embryo into an imaging analysis software, such as *ImageJ* that can handle time series data.

2. In *ImageJ*, select the ROI using one of the selection tools. For example, we used the elliptical tool to draw a circle around SAJs (Fig. 3d). The shape and dimensions of the ROI is kept constant for all subsequent measurements and will also be used to make background measurements.

3. Measure the fluorescence intensity of the ROI for each time point by first setting the measurements to "Mean Gray Value" (*Analyze>Set Measurements>Mean Grey Value*) and then measuring the intensity by hitting *Analyze>Measure* or *Ctrl-M*. Before performing fluorescence measurements, the image may need to be realigned if the region to be analyzed has moved over time. This can be done by manually repositioning the ROI, making the measurement then repeating these two steps for each time point or by automatically aligning the images prior to making the measurements using the *ImageJ* plugin TurboReg.

4. Import raw fluorescence measurements into a spreadsheet by copying and pasting.

5. Repeat measurements for 3–5 ROIs per embryo. Keep the shape and dimensions of the ROI constant.

6. Measure background (BKD) fluorescence intensity as above using the same sized selection areas as for the ROIs. Background readings are measured in regions of the image outside the embryo (Fig. 3d). Make three independent measurements per embryo. Import raw fluorescence measurements into the spreadsheet.

7. Measurements must also be made to correct for general acquisition photobleaching of the embryo while imaging. Measuring the fluorescence intensity of a group of multiple cells can be used to assess the rate of acquisition photobleaching (Fig. 3d). Measure the fluorescence intensity for the selected group of cells for each time point and import the raw data into the spreadsheet. This data will be referred to as "total intensity" and subsequently used when "correcting for total" described below. Keep the number of cells used for total intensity measurements consistent when analyzing different embryos.

8. All raw fluorescence intensity measurements should be arranged in columns in a single spreadsheet with the same time points for each row. A raw FRAP recovery curve can be generated by plotting fluorescence intensity versus time (Fig. 3e).

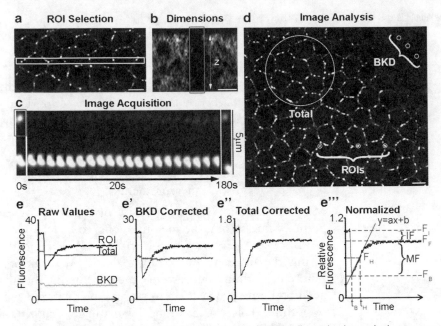

Fig. 3. Steps in measuring protein dynamics using FRAP. (**a**) A *box* of known dimension in *x–y* is drawn around the region to bleach. Here, a *rectangle* spanning five cells is drawn around SAJs in a late cellularization embryo expressing DE-cad::GFP. (**b**) Fixed embryos expressing tubulin::GFP were bleached revealing bleaching along the *z*-axis. An *x–z* view of a *z*-stack is shown. (**c**) Sample images of a FRAP experiment analyzing DE-cad::GFP recovery in SAJs at cellularization. The example shows prebleached at 0 s with the bleached region *boxed* and the unbleached SAJs below, recovery over 20 s (kymograph), and the recovery plateau at 180 s. (**d**) Regions of fluorescence intensity measurements used in generating fluorescence recovery curves. Shown here is an example of a prebleached image of an embryo expressing DE-cad::GFP. The fluorescence intensity of three regions of interest (ROI) was measured by drawing a circle around a single SAJ and measuring the "Mean Gray Value" in *ImageJ*. The same sized circles were used to make three background (BKD) intensity measurements outside the embryo. To correct for general acquisition photobleaching, the fluorescence intensity for a group of six cells was measured (Total). (**e**) Raw FRAP recovery curve. The fluorescence intensity within the selected areas is plotted versus time giving a raw FRAP recovery curve. (**e'**) A background corrected curve is obtained by subtracting the average BKD fluorescence intensity from the fluorescence intensity for the ROI at all time points. (**e''**) Next to correct for general acquisition photobleaching of the embryo, the BKD-subtracted intensity values are divided by total intensity values. (**e'''**) A normalized FRAP recovery curve is generated to make comparisons between experiments. The total-corrected data is normalized to one by dividing each time point by the fluorescence intensity of the time point immediately prior to bleaching. The data is now in relative fluorescence units. From this curve, multiple parameters can be calculated as outlined in "Graphical Representation and Calculating Descriptive Parameters" in Subheading 3.3.5. *IF* immobile fraction, *MF* mobile fraction, F_I initial fluorescence intensity, F_F final fluorescence intensity, F_B fluorescence intensity immediately after bleaching, F_H half of the final fluorescence intensity reached post-bleach, t_B time of bleaching, t_H time half of F_H. Bars = 5 μm.

Subtracting for Background Fluorescence

Background fluorescence results from the glass coverslip, objective and other sources. Average the three background measurements for each time point. Subtract this average background fluorescence intensity from the fluorescence intensity for the ROI for all time points (Fig. 3e'). These background corrected values are used in the next step.

Correcting for Intensity Changes Due to Acquisition Photobleaching and Laser Fluctuations

To correct for general bleaching of the embryo from imaging, divide background-subtracted intensity values by the total intensity values at each time point (Fig. 3e″).

Normalization of Data for Comparison of Different Experiments

To make comparisons between different experiments, the corrected fluorescence intensity measurements must be normalized. The initial fluorescence intensity or the time point immediately prior to bleaching is set to one and all other data points are normalized accordingly (Fig. 3e‴). The data is now in relative fluorescence units.

Graphical Representation and Calculating Descriptive Parameters

1. Make a plot of Time versus Relative Fluorescence. As shown in Fig. 3e‴, there is a pre-bleach plateau, a dip when bleaching was applied and then a recovery curve that reaches a second post-bleach plateau. From this plot, a number of parameters can be determined that describe the changes in the bleached and unbleached regions of the cell over time.

2. Immobile (IF) and mobile (MF) fractions of the fluorescent molecule of interest are determined from the recovery plot by examining the ratios of final fluorescence (F_F) reached post-bleach versus the initial fluorescence (F_I) intensity pre-bleach taking into account the fluorescence intensity immediately after bleaching (F_B) (Fig. 3e‴).

$$MF = (F_F - F_B) / (F_I - F_B) \text{ and } IF = 1 - MF.$$

To calculate the numbers of molecules in the immobile fraction, molecular numbers calculated in Subheading 3.2 can be applied.

3. The time required for half of the fluorescence molecules to exchange between the bleached and unbleached regions is referred to as the halftime of fluorescence recovery ($t_{1/2}$) and is commonly used to describe and compare FRAP experiments. Calculate half of the final fluorescence intensity reached post-bleach (F_H) and determine the time (t_H) this is achieved. Subtract the time of bleaching (t_B) and this is $t_{1/2}$ (Fig. 3e‴).

$$F_H = (F_F - F_B) / 2 \quad t_{1/2} = t_H - t_B.$$

4. To determine the rate of recovery, a best-fit line for the initial recovery period prior to reaching the post-bleach plateau is fitted and the slope calculated (see equation in Fig. 3e‴). This slope provides the rate of recovery as relative fluorescence intensity change per unit time. To calculate the number of molecules per unit time, molecule numbers calculated in Subheading 3.2 can be applied.

3.4. Statistical Considerations

In any experiment, results must be replicated to apply statistical tests. When analyzing cell biology in a developmental context, one must consider the populations of cells being analyzed and the populations of embryos being analyzed. Typically, an experiment is designed to probe for differences between one class of embryos and another (e.g. a mutant versus wildtype). Since embryos contain many cells, the total cell populations analyzed in an experiment can be very large. However, the total embryo populations may be relatively small. To test for differences between one class of embryos and another it is critical to apply a statistical test, such as a t-test, to data from the populations of embryos. To do this, we average the individual cell measurements from one embryo to provide one average value for that embryo. These values are then compared between the two populations of embryos with the statistical test. If the large populations of individual cell (or individual complex) measurements are compared, the statistical test may reveal a significant difference, but this may simply reflect natural variability between embryos rather than a specific difference between two different embryo populations.

4. Notes

1. Before beginning live imaging experiments the activity of the tagged protein should be assessed. Comparisons with immunofluorescence imaging of the endogenous protein are important for confirming that the tagged molecule is behaving normally. Addition of a GFP tag can alter protein localization and overexpression of tagged proteins can saturate endogenous protein localizing cues, leading to mislocalization. The tagged protein should also be examined in rescue experiments testing whether it can substitute for the endogenous protein.

2. The embryos are mounted with the vitelline membrane intact. The vitelline membrane is auto-fluorescent creating background noise. The yolk sack at the core of the embryo is also auto-fluorescent.

3. Image saturation blocks the resolution of discrete structures (Fig. 1c) and prevents quantification of signal levels.

4. The subtraction of the cytoplasmic profiles assumes there is zero cytoplasmic protein localization. Thus, protein counts at the cell cortex will likely be overestimates.

5. The column containing the glutathione resin will change from white to light green when the GST-GFP is bound.

Acknowledgements

Work in our lab is supported by a CIHR operating grant and an NSERC operating grant. A. McKinley holds an Ontario Graduate Scholarship in Science and Technology. T. Harris holds a Tier 2 Canada Research Chair.

References

1. McGill MA, McKinley RF, Harris TJ (2009) Independent cadherin-catenin and Bazooka clusters interact to assemble adherens junctions. *J Cell Biol*. 185:787–796.

2. Pope KL, Harris TJ (2008) Control of cell flattening and junctional remodeling during squamous epithelial morphogenesis in Drosophila. *Development*. 135:2227–2238.

3. David DJ, Tishkina A, Harris TJ (2010) The PAR complex regulates pulsed actomyosin contractions during amnioserosa apical constriction in Drosophila. *Development*. 137:1645–1655.

4. Wu JQ, Pollard TD (2005) Counting cytokinesis proteins globally and locally in fission yeast. *Science*. 310: 310–314.

5. Wu JQ, McCormick CD, Pollard TD (2008) Chapter 9: Counting proteins in living cells by quantitative fluorescence microscopy with internal standards. *Methods Cell Biol*. 89:253–273.

6. Fowlkes CC, Hendriks CL, Keranen, SV et al (2008) A quantitative spatiotemporal atlas of gene expression in the Drosophila blastoderm. *Cell*. 133:364–374.

7. Lippincott-Schwartz J, Snapp E, Kenworthy A (2001) Studying protein dynamics in living cells. *Nat Rev Mol Cell Biol*. 2:444–456.

8. Lippincott-Schwartz J, Altan-Bonnet N, Patterson GH (2003) Photobleaching and photoactivation: following protein dynamics in living cells. *Nat Cell Biol*. Suppl: 5:S7–14.

Chapter 2

Analyzing Frizzled Signaling Using Fixed and Live Imaging of the Asymmetric Cell Division of the *Drosophila* Sensory Organ Precursor Cell

Bertrand Jauffred and Yohanns Bellaiche

Abstract

When you look at the dorsal thorax of a fruitfly, you can easily get fascinated by the high degree of alignment of the bristles that show a strong polarization in their surface organization. This organization of cells in the plane of the epithelium is known as planar cell polarity (PCP), and was initially characterized in *Drosophila melanogaster*. This process is important in a broad variety of morphological cellular asymmetries in various organisms. In *Drosophila*, genetic studies of PCP mutants showed that the asymmetric division of the sensory organ precursor cell (pI cell) is polarized along the anterior–posterior axis by Frizzled receptor signaling. Here, we described two methods to image and analyze the PCP in the pI cell model.

Key words: Asymmetric division, Sensory organ precursor cell (SOP or pI cell), Notum, Dissection, Confocal imaging

1. Introduction

During the development of metazoans, several cell types are generated. Asymmetric divisions are one of the mechanisms leading to this cell diversity (1). The unequal inheritance of the cell fate determinants by the two daughter cells requires a spatial and temporal coordination between the positioning of the mitotic spindle and the asymmetric localization of the cellular determinants. The identification and the characterization of the cell determinants were initially made in the fruitfly *Drosophila melanogaster* and the worm *Caenorhabditis elegans*. The asymmetric cell division of the *Drosophila* sensory organ precursor (SOP) cell has been used a central paradigm to understand how the Frizzled planar cell polarity (PCP) pathway polarizes asymmetric cell division (2).

Kursad Turksen (ed.), *Planar Cell Polarity: Methods and Protocols*, Methods in Molecular Biology, vol. 839,
DOI 10.1007/978-1-61779-510-7_2, © Springer Science+Business Media, LLC 2012

The *Drosophila* dorsal thorax or notum consists of 100 peripheral sensory organs laid out in lines. Each sensory organ is made of four cells, which are produced by a succession of asymmetric divisions from only one SOP cell, the pI cell (2). pI cells are specified in the epithelium and divide asymmetrically within the plane of the epithelium, along the anterior–posterior axis in a Frizzled-dependent manner. During the metaphase of the first asymmetric division, the cell determinants, Numb (3) and Neuralized (4) localize at the anterior cortex and the mitotic spindle aligns with the anterior–posterior axis. Hence, Numb and Neuralized segregate into the only anterior pIIb cell (5). In the pIIb cell, Numb and Neuralized control the endocytosis of Notch and the recycling of Delta, respectively (6). The asymmetric division of pI cell is a powerful system to study in vivo cell polarization in response to Frizzled signaling by using confocal microscopy on fixed tissue, or time-lapse confocal microscopy on living organisms.

2. Materials

2.1. Dissection and Mounting of the Drosophila Notum (Fig. 1a)

- Dumont #5 Forceps – Dumostar Biologie; Fine Science Tools (FST #11295-10).

- Cohan-Vannas Spring Scissors – 5-mm Blade Straight Sharp; Fine Science Tools (FST #15000-02) or Vannas Scissors – 5-mm Blade Straight Sharp; Moria Surgical (#9600).

- Glass Pasteur Pipettes 150 mm.

- Stainless Stell Minutiens Pins – 0.1 mm Diameter; Fine Science Tools (FST #26002-10).

- Tissue culture dishes 40 × 11 mm.

- Kit Rhodrosil® RTV-2 Silicones; Bluestar Silicones.

- Bulb to pipette.

- Glass watch.

- Phosphate-buffered saline (PBS): Prepare 10× stock with 1.37 M NaCl, 27 mM KCl, 100 mM Na_2HPO_4, 18 mM KH_2PO_4 (adjust to pH 7.4 with HCl if necessary) and autoclave before storage at room temperature. Prepare working solution (PBS1×) by dilution of one part with nine parts of water.

- PBS with 0.1% Triton X-100 (SIGMA 9002-93-1) (PBT): Dilute 1 mL with 1 L PBS1×.

- PBS saline with glycerol: Prepare a 50% (v/v) solution of glycerol in PBS1.

- Paraformaldehyde (Electron Microscopy Science): Prepare a 4% (v/v) solution in PBS10× by dilution of one ampoule of paraformaldehyde aqueous solution 16% (10 mL) with 4 mL PBS10× and 26 mL water. The solution may need to be

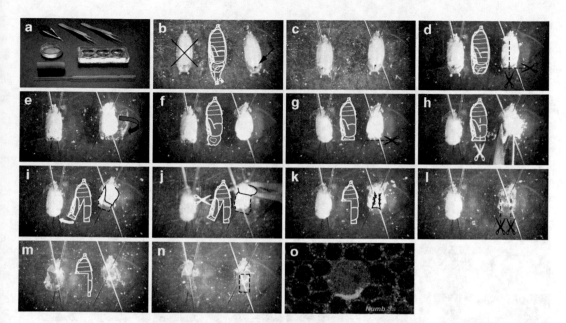

Fig. 1. Sequence of the notum dissection. The pupae are well positioned and pinned down. After that, all the steps of the dissection occur in PBS1×. The aim of the dissection is the individualization of notum to microscopy imaging.

carefully manipulated under a hood. Aliquote in 2 mL tubes and store at −20 °C, stable for 2 months.

– Mounting medium with DAPI: Prepare a 20% (v/v) PBS1×, 80% (v/v) glycerol, 2% (w/v) N-propylgalate, and 1:2,000 (v/v) DAPI.

2.2. Dissection and Mounting of the Drosophila Pupae Between Slide and Cover Slip (Fig. 2a)

– Dumont #5 Forceps – Dumostar Biologie; Fine Science Tools (FST #11295-10).

– Microscope slides 25 × 75 × 1.0 mm; Thermo Scientific.

– Cover glass 24 × 40 mm st.1; Knittel Deckgläser.

– Cover slips 18 × 18 mm no.1; Knittel Deckgläser.

– Voltalef oil 10S; VWR Prolabo (#24 627.188).

– Double sided tape; Scotch® 3M.

– Nail polish fast dry.

– Kleenex® paper towel.

3. Methods

3.1. Dissection

Flies were cultured on standard yeast–cornmeal–sugar medium at 25°C. Flies of the genotype w^{1118} were used as wild type. Pupae were collected at the puparium formation and transferred into small Petri dishes until the dissections of the nota were processed

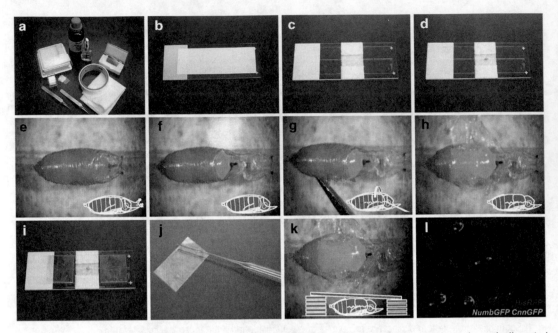

Fig. 2. Sequence of pupa mounting between slide and cover slip. The pupa is glued and the dorsal pupal case is dissected. The aim of the pupae dissection is the analysis of notum by time-lapse confocal microscopy.

at 16–17 h after puparium formation (APF). The entire dissection should not exceed 15 min before fixation.

1. Place the pupae on the dorsal side (mouth hooks facing upwards) in a Petri dish half filled with polymerized RTV silicone (Fig. 1b).

2. Pin the pupae down with manutien pins which should be positioned in the rear half of the abdomen through the pupal case (Fig. 1c).

3. Add sufficient PBS1× to submerge the pupae.

4. Cut the spiracles (respiratory organs) anterior to the head of the pupae with microscissors (Fig. 1d).

5. Cut the pupal case longitudinally up to the abdomen (Fig. 1d).

6. Fold back each pupal case part with forceps in order to remove it (Fig. 1e).

7. Cut the pupae in the middle of the head (Fig. 1g).

8. Insert the scissors in the opening to cut the lateral side of the pupae up to the abdomen (Fig. 1h).

9. Repeat the cut on the other side.

10. Separate both parts and remove the fat bodies by pipetting gently (Fig. 1i).

11. Cut and remove the ventral part (Fig. 1j).

12. Remove the remaining fat bodies by pipetting up and down gently.

13. Remove the trachea (Fig. 1k).

14. Cut the notum to create a rectangular shape (Fig. 1l) (you can perform this step after fixation).

15. Make different notches (rectangular, triangular, ...) in head part. Notches of different types can be cut to distinguish between genotypes (see Note 1).

16. Remove the PBS1× and immediately fix the dissected nota for 20 min by adding a 4% paraformaldehyde, PBS1× solution. The fixation solution may need to be carefully manipulated under a hood.

17. Remove the fixation solution and rinse three times with PBS1×, 0.1% Triton X-100 solution (PBT); ensuring that the nota does not dry!

18. Incubate at least 20 min in PBT.

3.2. Immunohisto-chemistry

For the all next steps, incubations occur at room temperature in a "humid chamber" to avoid dehydrating the nota.

1. Isolate each notum by cutting at the junction between the scutellum and the abdomen and transfer them to a three-well glass dish filled with PBT.

2. Remove PBT and add the primary antibody solution to the well.

3. Incubate for 90 min.

4. Recuperate the primary antibody solution for future use and add PBT to the well.

5. Perform three 10-min washes in PBT.

6. Remove PBT and add the secondary antibody solution to the well.

7. Incubate for 30 min.

8. Remove and discard the secondary antibody solution and add PBT to the well.

9. Perform three 10-min washes in PBT.

10. Remove PBT and add 50% PBS, 50% glycerol (v/v) solution.

11. Incubate for at least 45 min at room temperature or at 4°C overnight.

3.3. Mounting of Dissected Nota

1. Put a drop of mounting medium with DAPI on a slide.

2. Transfer the nota one by one from three-well glass dish filled with 50% PBS, 50% glycerol (v/v) solution to the mounting medium drop.

3. Orient the nota with apical domain up (cuticle is glossy whereas the epithelium is matte) and align them to facilitate the imaging (see Note 1).

4. Lay the cover slip down gently and stabilize it with nail polish.

5. Store at 4°C.

3.4. Mounting of Living Pupae

Flies were cultured on standard yeast–cornmeal–sugar medium at 25°C. Flies of the genotype neur-Gal4, UAS-numb::GFP, UAS-cnn::GFP, UAS-histone2B::mRFP were used. Pupae were collected at the puparium formation and transferred into small Petri dishes until the mounting of the pupae were processed at 16–17 h APF.

1. Clean the pupae with a damp paper towel (see Note 2).

2. Glue the ventral side of the pupae on the slide with double-sided tape (Fig. 2b). Wrap damp paper towel around the slide (see Fig. 2c) to humidify the sample (Fig. 2d).

3. Gently remove the opercula by pulling a spiracle (Fig. 2f).

4. Insert forceps at opercula opening and tear laterally from head to abdomen to remove the pupal case (Fig. 2g) (see Note 3).

5. Fold back the pupal case on the other side with forceps (Fig. 2h).

6. Create stacks composed of four 18×18 mm cover slips for the front stack and five 18×18 mm cover slips for the rear stack. Use nail polish to glue cover slips together and adhere stacks to slide using double-sided tape (Fig. 2i).

7. Coat a 24×40 mm cover slip with Voltalef 10S oil on its contact side to increase optical resolution (Fig. 2j), onto the pupae and the backings. Cement it with nail polish. The contact between the notum and the oil must create a meniscus (Fig. 2k).

4. Notes

1. Play within the light system permits increase of the contrast between the nota and the environment.

2. Cleaning the pupae improves adhesion on the double-sided tape, by removing the food left on pupal case.

3. Using forceps with rounded tips can be less wounding for the pupae.

Acknowledgments

The authors would like to thank Rachel Duffie and Boris Guirao for critical reading. Work in Bellaiche lab is supported by grants to Y.B. from the Association pour la Recherche sur le Cancer (ARC 4830), the ANR (BLAN07-3-207540), the CNRS, INSERM, ERC Starting Grant (CePoDro 209718), and the Curie Institute.

References

1. Betschinger, J., Knoblich, J.A. (2004) Dare to be different: asymmetric cell division in Drosophila, C. elegans and vertebrates. *Curr Biol.* **14,** R674–85.

2. Gho, M., Bellaïche, Y., and Schweisguth, F. (1999). Revisiting the Drosophila microchaete lineage: a novel intrinsically asymmetric cell division generates a glial cell. *Development.* **126,** 3573–3584.

3. Rhyu, M.S., Jan, L.Y. & Jan, Y.N. (1994) Asymmetric distribution of numb protein during division of the sensory organ precursor cell confers distinct fates to daughter cells. *Cell* **76,** 477–491.

4. Le Borgne, R., Schweisguth, F. (2003) Unequal segregation of Neuralized biases Notch activation during asymmetric cell division. *Dev Cell.* **5,** 139–148.

5. Bellaïche, Y., Gho, M., Kaltschmidt, J. A., Brand, A. H. and Schweisguth, F. (2001). Frizzled regulates localization of cell-fate determinants and mitotic spindle rotation during asymmetric cell division. *Nat. Cell Biol.* **3,** 50–57.

6. Le Borgne, R., Bardin, A., Schweisguth F. (2005) The roles of receptor and ligand endocytosis in regulating Notch signaling. *Development.* **132,** 1751–1762.

Chapter 3

Protein–Protein Interaction Techniques: Dissect PCP Signaling in *Xenopus*

Yingqun Wang

Abstract

The planar cell polarity (PCP) pathway is a β-catenin-independent branch of the Wnt signaling cascade. In vertebrate embryos PCP signaling regulates morphogenetic events including convergent extension (CE) movements during gastrualtion. *Xenopus* embryo has been established as an excellent model system to dissect PCP signaling in vertebrates because morphogenetic cell behaviors including CE can easily be monitored in vivo. *Xenopus* Paraxial protocadherin (xPAPC) is a transmembrane protein which serves as a link between patterning factors in the Spemann's organizer and regulators of the morphogenetic movements. xPAPC regulates morphogenesis in part by modulating cell adhesion and PCP signaling. Here two methods, GST pull-down assay and yeast two-hybrid assay, are described for the identification of xPAPC interacting proteins to elucidate the mechanism by which xPAPC regulates PCP signaling.

Key words: Wnt, Planar cell polarity, *Xenopus*, Protein–protein interaction, GST pull-down, Yeast two-hybrid, Paraxial protocadherin

1. Introduction

Morphogenesis is the key process that generates tissue organization and shape during organism development. Over the past several decades, a wide variety of molecules and pathways have been identified that are implicated in the regulation of morphogenetic cell behaviors. Wnt/planar cell polarity (PCP) pathway has emerged as an important player in morphogenesis in various organisms, ranging from the precise arrangement of hairs and bristles in *Drosophila*, convergent extension (CE) movements of mesenchymal cells in *Xenopus*, to the stereociliary bundle orientation in the cochlea in mammals (1–3). In addition, abnormal PCP signaling has been involved in human diseases such as developmental defects

Kursad Turksen (ed.), *Planar Cell Polarity: Methods and Protocols*, Methods in Molecular Biology, vol. 839,
DOI 10.1007/978-1-61779-510-7_3, © Springer Science+Business Media, LLC 2012

and cancer (3, 4). Therefore, it is of fundamental importance to provide mechanistic insights into PCP signaling pathway, especially in vertebrates.

In *Xenopus* embryos, PCP signaling can be studied with relative ease due to the obvious convergent extension (CE) phenotype, which is delicately modulated by PCP signaling during gastrulation. In *Xenopus*, CE movements are considered as the main driving force of gastrulation, a crucial step in early embryogenesis that contributes to the establishment of the basic body plan and formation of primary germ layers. Substantial progress has been made recently to identify and characterize pathways and molecules implicated in the modulation of CE movements during *Xenopus* gastrulation (5). *Xenopus* Paraxial protocadherin (xPAPC) is an important molecule that links regulatory genes in the Spemann's organizer with the execution of morphogenesis including CE movements in *Xenopus*. Moreover, the regulation of morphogenesis by xPAPC depends on its ability to modulate the activity of Rho GTPase and c-jun N-terminal kinase, which are effectors of the PCP pathway (6, 7). To elucidate the detailed mechanism by which xPAPC regulates PCP signaling, it is imperative to identify xPAPC interacting partners that may provide the missing link between xPAPC and PCP signaling.

A variety of methods have been developed to detect protein–protein interaction since protein–protein interactions are engaged in nearly all biological processes. Each method has its own advantages and limitations, especially with regard to the sensitivity and specificity of the methods. Co-immunoprecipitation is considered to be the gold standard assay for protein–protein interactions, especially when it is performed with endogenous proteins. However, this method can only verify interactions between suspected interaction partners and is not a screening approach. Yeast two-hybrid method has high sensitivity but is subjected to a notorious high false-positive rate. On the other hand, the intrinsic nature of particular proteins will dictate, to a large extent, which approaches are suitable for identification of their interacting partners. For instance, if an interaction is likely to depend on posttranslational modifications that occur only in eukaryotes, GST pull-down will not be appropriate. If a protein such as transcription factor can autoactivate transcription of reporter genes, it is not suitable for yeast two-hybrid assay. Moreover, if a protein interacts simultaneously with multiple partners in a complex fashion, yeast two-hybrid may fail to identify these partners. Therefore, it is best to combine different approaches to identify interacting partners of a particular protein.

Here, we present the methodology details of GST pull-down assay and yeast two-hybrid assay to identify xPAPC interacting proteins, demonstrating an example of how to dissect PCP signaling in *Xenopus*. Using these methods we successfully identified Sprouty as a new binding partner of xPAPC and established that xPAPC promotes gastrulation movements by sequestration of Sprouty, thus unraveling a novel mechanism by which protocadherins modulate PCP signaling (8).

2. Materials

2.1. Stock Solutions, Media, and Buffers for Yeast Work

1. Single-stranded carrier DNA: Weigh 200 mg of salmon sperm DNA Type III Sodium Salt (Sigma D1626) into 100 ml of TE buffer (pH 8.0) or ddH$_2$O. Disperse the DNA into solution by drawing it up and down repeatedly with a 25-ml plastic pipette. Mix vigorously on a magnetic stirrer for 2–3 h or o/n in a cold room. Aliquot DNA in 1 ml and store at –20°C.

2. 1 M LiOAc: Adjust to pH 7.5 with dilute acetic acid and autoclave.

3. 50% PEG (w/v): Weigh 50 g PEG (MW 3350, Sigma P3640) into a beaker and add 80 ml ddH$_2$O. Stir with a magnetic stirring bar until dissolved. Transfer all of the liquid to a 100-ml graduated cylinder. Wash beaker with a small amount of ddH$_2$O and add this to the graduated cylinder containing the PEG solution, and bring the volume to exactly 100 ml. Mix well by inversion. Autoclave at 121°C for 15 min.

4. 1 M 3-AT: prepare in ddH$_2$O and filter sterilize. Store at 4°C. Store plates containing 3-AT at 4°C for up to 2 months.

5. X-Gal (20 mg/ml in DMF): dissolve X-Gal in DMF. Store in the dark at –20°C.

6. 40% Galactose: dissolve 40 g galactose in 100 ml H$_2$O, autoclave at 121°C for 15 min.

7. 40% Raffinose: dissolve 40 g raffinose in 100 ml H$_2$O, autoclave at 121°C for 15 min.

8. 10× BU salts: 70 g/l Na$_2$HPO$_4$·7H$_2$O, 30 g/l NaH$_2$PO$_4$, add H$_2$O to 1 l, dissolve and autoclave, and store at RT.

9. 5× M9 Salts: 64 g/l Na$_2$HPO$_4$·7H$_2$O, 15 g/l KH$_2$PO$_4$, 2.5 g/l NaCl, 5 g/l NH$_4$Cl, add H$_2$O to 1 l, and autoclave at 121°C for 15 min.

10. YPAD media: 50 g/l YPD powder, 0.03 g/l adenine hemisulfate, 20 g/l agar (for plates only). Add H$_2$O to 1 l, autoclave at 121°C for 15 min.

11. 2× YPAD media: 100 g/l YPD powder, 0.03 g/l adenine hemisulfate. Add H$_2$O to 1 l, autoclave at 121°C for 15 min.

12. SD media: 27 g/l DOB (dropout base) powder, CSM-Amino acid (Complete Supplement Mixture minus the above amino acid), 0.03 g/l adenine hemisulfate, 20 g/l agar (for plates only). Add H$_2$O to 1 l, dissolve, and autoclave at 121°C for

15 min. For SD plates containing 3-AT, cool SD media to 55°C and add appropriate amount of 1 M 3-AT stock solution and swirl to mix well before pouring to plates.

13. SD-Trp/-Leu/-His/+10 mM 3-AT/+X-Gal plate: 27 g/l DOB powder, 0.62 g/l CSM-Trp-Leu-His, 0.03 g/l adenine hemisulfate, 20 g/l agar. Add H_2O to 820 ml, dissolve, and autoclave at 121°C for 15 min. Cool to 55°C, add 50 ml 40% galactose, 25 ml 40% raffinose, 100 ml 10× BU salts, 4 ml X-Gal stock, 10 ml 1 M 3-AT stock. Pour and store plates inverted in the dark at 4°C.

14. M9/Amp minimal medium plates (for selection of prey plasmid in KC8 cells): 0.69 g/l CSM-Leu, 20 g/l agar, 750 ml H_2O. Autoclave at 121°C for 15 min. Cool to 55°C, add 200 ml 5× M9 salts, 20 ml 20% glucose, 2 ml 1 M $MgSO_4$, 0.1 ml 1 M $CaCl_2$, 1 ml 1 M thiamine HCl (filter-sterilized), 1 ml 50 mg/ml ampicillin. Pour and store plates at 4°C.

15. Buffers for β-galactosidase filter assay.

Z buffer: $Na_2HPO_4 \cdot 7H_2O$ 16.1 g/l, $NaH_2PO_4 \cdot H_2O$ 5.50 g/l, KCl 0.75 g/l, $MgSO_4 \cdot 7H_2O$ 0.246 g/l. Adjust to pH 7.0 and autoclave.

X-gal stock solution: Dissolve X-Gal in DMF at a concentration of 20 mg/ml. Store in the dark at –20°C.

Z buffer/X-gal solution: 100 ml Z buffer, 0.27 ml β-ME, 1.67 ml X-gal stock solution.

16. Buffers for yeast protein extraction.

NaOH/β-ME buffer: 1.85 M NaOH, 7.5% β-mercaptoethanol.

55% TCA (w/v): To new bottle containing 500 g of TCA, add 227 ml H_2O to get 100% TCA (w/v). Then mix 55 ml to 45 ml H_2O.

SU buffer: 5% SDS (w/v), 8 M urea, 125 mM Tris–HCl (pH 6.8), 0.1 mM EDTA, 0.005% bromophenol blue (w/v). Store at –20°C. Add 15 mg of dithiothreitol/ml of SU buffer prior to use.

2.2. Media and Buffers for GST Pull-Down Assay

1. 2×YTA medium: 16 g/l tryptone, 10 g/l yeast extract, 5 g/l NaCl, 100 μg/ml ampicillin, pH 7.0.

2. MBSH buffer: 88 mM NaCl, 1 mM KCl, 2.4 mM $NaHCO_3$, 0.8 mM $MgSO_4$, 0.33 mM $NaNO_3$, 0.4 mM $CaCl_2$, 10 mM HEPES pH 7.4, 10 μg/ml streptomycin, 10 μg/ml penicillin.

3. PBS lysis buffer: 20 mg lysozyme, 40 μl 0.5 M EDTA, 200 μl 100 mM PMSF, 200 U DNase I in PBS.

4. Buffer H: 50 mM Tris–HCl, pH 7.5, 250 mM KCl, 5 mM EDTA, 1 mM dithiothreitol, 0.1% Triton X-100.

2.3. Bacterial, Yeast Strains, and Yeast Two-Hybrid cDNA Library

1. *E. coli* XL1-Blue: *recA1 endA1 gyrA96 thi-1 hsdR17 supE44 relA1 lac* [F′ *proAB lacI*ᵠZΔM15 Tn*10* (Tetʳ)].

2. *E. coli* BL21: F⁻ *ompT hsdS*$_B$(r$_B$⁻ m$_B$⁻) *gal dcm*.

3. *Saccharomyces cerevisiae* L40: *MAT*a *ade2 his3 leu2 trp1 LYS2::lexA-HIS3 URA3::lexA-lacZ*.

4. *Xenopus laevis* oocyte Matchmaker cDNA library (Clontech): kindly provided by Staub (9) as frozen glycerol stock of *E. coli* transformed with the library.

2.4. Oligonucleotides

The sequences of all oligonucleotides are in 5′–3′ direction as specified.

For sequencing

pACT2-derived plasmids:

pACT2-U2: GTGAACTTGCGGGGGTTTTTCAGTATCTACG

pACT2-D2: ATACGATGTTCCAGATTACGCTAGCTTGGG

For cloning (restriction site underlined)

Clone xPAPCc into pNLX3:

Forward: CGC<u>GGATCC</u>GTTGTACTTGTAAAAAGAAAG

Reverse: CGT<u>CTGCAG</u>AAAGGTTGTAGCAATTTCTG

Clone xPAPCc into pGEX6p2:

Forward: GGC<u>GTCGAC</u>GTGTACTTGTAAAAAGAAAGC

Reverse: AT<u>GCGGCCGC</u>AAAGGTTGTAGCAATTTCTG

2.5. Plasmids

1. Plasmids for yeast two-hybrid assay: Empty bait plasmid pNLX3 was provided by Moreau (10) and prey plasmid pACT2 by Staub O. cDNA for the cytoplasmic domain of xPAPC (residues 715–979) was amplified by PCR and cloned into the pNLX3 using *Bam*HI/*Pst*I sites in frame with LexA BD domain to make the construct pNLX3–xPAPCc.

2. Plasmids for GST pull-down assay: cDNA for the cytoplasmic domain of xPAPC was amplified by PCR and cloned into the pGEX6p2 using *Sal*I/*Not*I sites to make the construct pGEX6p2–xPAPCc.

3. Methods

3.1. Preparation of Xenopus laevis embryos

1. Chorionic gonadotropin (see Note 1) dissolved in ddH$_2$O were subcutaneously injected to the dorsal lymph sac of healthy female frogs. Generally, the injected frogs were kept at RT waterbath o/n and began spawning eggs the next morning.

2. Testes used for in vitro fertilization were dissected from adult male frogs and maintained in 1× MBSH buffer at 4°C, which are viable of fertilization for at least 2 weeks.

3. For in vitro fertilization, female frogs were gently squeezed to lay eggs into a large petri dish and excess liquid was removed. Subsequently, a small piece of testis was cut into fine pieces with scissors, suspended in 0.5–1 ml 1× MBSH and transferred onto the eggs. The eggs were gently mixed well with the sperm suspension, spread to a single layer on the bottom of petri dish. Then plenty volume of H_2O was added to cover the fertilized eggs. About 30 min later, fertilization rates can be determined by observing cortical rotation. Embryos were dejellyed with 2% cysteine (see Note 2), washed intensively with water, and cultured in 1× or 0.1× MBSH depending on the desired stages.

3.2. SDS-PAGE and Western Blot

1. SDS-polyacrylamide gel electrophoresis: Separation of proteins was performed by means of the discontinuous SDS-polyacrylamide gel electrophoresis (SDS-PAGE). The size of the running and stacking gel was as follows: Separation gel: height ~8 cm, thickness 1 mm, 10%(v/v) acrylamide; Stacking gel: height ~2 cm, thickness 1 mm, 5% (v/v) acrylamide, 10- or 15-well combs. After complete polymerization of the gel, the chamber was assembled as described by the manufacturer's protocol. Samples were loaded in the pockets and the gel was run at constant current at 10 mA for the stacking gel and then for the separation gel at 20 mA. The gel run was stopped when the bromophenol blue dye had reached the end of the gel. Gels were then either stained or subjected to Western Blotting.

2. Western Blot: Proteins were transferred from the SDS-gel onto a nitrocellulose membrane (Amersham) using wet transfer method. Proteins were transferred on ice at 400 mA for 90 min. The prestained protein marker was used as molecular weight marker and to monitor the transfer. After electrophoretic transfer, the membranes were removed from the sandwich and placed in plastic dishes. Membranes were washed once in PBS-T and then incubated in blocking buffer for 30 min at RT with gently shaking. Afterward, the primary antibody was added in an appropriate dilution and incubated either for 1 h at RT or o/n at 4°C. The primary antibody was removed and the membrane was washed 3× 10 min with PBS-T. The appropriate secondary antibody was then applied for 1 h at RT. The membrane was washed again 3× 10 min with PBS-T. The antibody bound to the membrane was detected by using the enhanced chemiluminescence detection reagent. The membrane was incubated for 2–3 min in detection reagent. Then the solution was removed and the blot was dried and placed between two saran warp foils. The membrane was exposed to X-ray film (Biomax-MR, Kodak) for several time periods and the films were developed in a dark room in an automatic developing machine.

3. Coomassie staining of polyacrylamide gels: The coomassie staining of polyacrylamide gels was performed with the Imperial Protein Stain kit (Pierce) following the manufacturer's protocol.

3.3. GST Pull-Down Assay

3.3.1. Expression of GST-Fusion Protein

1. Transform recombinant GST expression vector into *E. coli* strain BL21. Inoculate single recombinant colony into 4 ml 2×YTA medium (see Note 3). Or inoculate 40 μl o/n culture into 4 ml medium (100-fold dilution).

2. Grow to OD_{600} of 0.6–0.8 with vigorous agitation at 37°C (3–5 h).

3. Divide into four tubes and add 10 μl 100 mM IPTG (1 mM final concentration) into three tubes.

4. Continue incubation for additional 2, 4, and 6 h, respectively.

5. Harvest cells by centrifuging 1 ml culture.

6. Resuspend each pellet in 40 μl 1× SDS loading buffer, heat at 100°C for 3 min and spin.

7. Load 20 μl sample and run 10% SDS-PAGE.

8. Stain gel with Coomassie Brilliant Blue to check which colony express GST-fusion protein.

3.3.2. Batch Purification of GST-Fusion Protein

1. Inoculate positive colony in 20 ml 2×YTA medium, 37°C o/n.

2. Dilute into 500 ml 2×YTA medium, 37°C till $OD_{600} = 0.8$.

3. Induce by adding 100 mM IPTG to final concentration of 1 mM.

4. Incubate for additional 3 h at 30°C (see Note 4).

5. Spin and resuspend in 20 ml PBS lysis buffer, stir at RT for 30 min.

6. Optional: the cells can be disrupted by sonication (see Note 5).

7. Add Triton X-100 to 1% and dithiothreitol to 15 mM, mix (see Note 6).

8. Spin at $13,000 \times g$ at 4°C for 30 min.

9. Take 1.33 ml Glutathione Sepharose 4B (Amersham Pharmacia Biotech, Freiburg), spin, and wash with 10 ml PBS twice and resuspend to 1 ml in PBS.

10. Mix the supernatant from step 8 with 1 ml beads from step 9, incubate at 4°C for 1 h.

11. Spin and wash 3–4 times with 10 ml ice-cold PBS (+1 mM PMSF and 15 mM dithiothreitol).

12. The beads with the bound GST or GST-fusion protein are kept for 4°C and ready for use with GST pull-down. For long-term storage, add 10% glycerin to the beads and keep at −20°C (see Note 7).

3.3.3. GST Pull-Down of Embryo Extracts

1. Gastrulation stage *Xenopus* embryos were homogenized in buffer H with protease inhibitors. 100 embryos/ml buffer H.

2. The extracts were centrifuged at $13,000 \times g$ at 4°C for 30 min. The supernatant was taken and centrifuge again at $13,000 \times g$ at 4°C for 30 min (see Note 8).

3. The supernatant was cleared by incubation with GST Sepharose beads o/n at 4°C.

4. After centrifugation at $13,000 \times g$ at 4°C for 30 min, the precleared supernatant was incubated o/n with GST-xPAPCc or GST Sepharose beads at 4°C.

5. The beads were washed three times with buffer H. The bound proteins were released by boiling in SDS loading buffer and resolved by 10% SDS-PAGE.

6. For mass spectrometry, the gels were subjected to coomassie staining with the Imperial Protein Stain kit. The bands of interest were excised from the gels and sent for mass spectrometry protein identification in Center for Molecular Biology, University of Heidelberg.

3.4. Yeast Two-Hybrid Assay

3.4.1. Preparation of Frozen Competent L40 Yeast Cells

1. Grow L40 stain in appropriate media (YPAD) 30° o/n. Approximately 1 ml for every 100 ml intended to inoculate the next day (i.e., 2.5 ml o/n to inoculate 250 ml). For a yeast strain with a selectable plasmid, grow in appropriate selection media about 20 ml for every 100 ml intended to inoculate the next day (see Note 9). Some yeast strains in selection media may take 2–3 days to reach saturation. Grow yeast just to the point of saturation ($OD_{600} = 1.0$).

2. Inoculate 250 ml YPAD this will be enough for about every seventy-five 100 µl aliquots of frozen competent cells. Scale-up as desired. Let them grow to log phase ($OD_{600} \sim 0.7$). For a strain with a selectable marker inoculate to $OD_{600} \sim 0.3$ so it goes through one doubling time (see Note 10).

3. Spin down cells ($6,000 \times g$ for 10 min) and wash in 0.4 volumes of starting volume (i.e., 40 ml for 100 ml culture) of 100 mM LioAC.

4. Spin down cells again and wash in a 0.2 volumes starting with 100 mM LioAC.

5. Spin down a final time and resuspend cells in 100 mM LioAC with 15% glycerol to a final volume of 0.03 of starting volume.

6. Aliquot 100 µl shots in microfuge tubes and put cells into a cardboard box and allow to freeze slowly in –80°C (see Note 11).

3.4.2. Transformation of Frozen Yeast Cells

1. Spin down 100 µl frozen competent cells from –80°C (or made fresh) 1 min at $\sim 14,000 \times g$. Use one tube per single or double transformation.

2. Aspirate the supernatant and add the following in order. 50% PEG-240 µl, 1 M LiAc-36 µl, ssDNA-79 µl, plasmid DNA-5 µl (about 200 ng), total-360 µl (see Note 12). If doing a double transformation use 3 µl of each of the plasmid DNA and reduce the amount of ssDNA to 78 µl per transformation to keep the DNA/PEG-LiAc ratio the same.

3. Resuspend the yeast cells in the mixture by vortexing well to remove any clumps.

4. Incubate on rocker in 30°C for 30 min at 150 rpm.

5. Heat shock in 42°C water bath for 15 min.

6. Spin down at 14,000×*g* for 1 min., aspirate off the supernatant.

7. Resuspend cells in 200 μl ddH$_2$O and plate all cells on appropriate selection media plates.

8. Incubate plates in 30°C for 2–4 days until colonies appear.

3.4.3. Pretransformation of L40 with xPAPCc Bait Plasmid

1. According to method in Subheading 3.4.2, transform xPAPCc bait plasmid, pNLX3–xPAPCc, into L40 and plate on SD-Trp plate.

2. Inoculate the positive clones in 5 ml SD-Trp media and cultured at 30° till saturation (OD$_{600}$ > 1.0).

3. Pellet the yeast cells by spinning the culture at 1,000×*g* for 5 min at RT.

4. Resuspend the pellet with 1 ml of cold ddH$_2$O and add 150 μl of fresh-made NaOH/β-ME buffer.

5. Vortex the cells for 30 s and incubate on ice for 15 min.

6. Vortex again and add 150 μl of 55% TCA (in water). Vortex and place on ice for 10 min.

7. Collect the protein extracts by centrifugation at 12,000×*g* for 10 min at 4°C. Remove the supernatant and centrifuge again to remove any residual supernatant.

8. Resuspend the pellet in 300 μl of SU buffer. Add 1–2 μl of Tris–HCl (pH 8.0) if the solution turns yellow. Vortex to resuspend the protein pellet. Heat at 65°C for 3 min prior to loading 30–40 μl sample for SDS/PAGE.

9. Analyze by standard Western Blot method using antibody against LexA. The positive clones that express LexA-bait fusion protein are called L40-xPAPCc.

3.4.4. Check Auto-activation of LacZ Reporter Gene in L40-xPAPCc

1. Streak L40-xPAPCc strain on SD-Trp + X-Gal plates and incubate at 30°C for 2–4 days.

2. Check whether the colonies turn blue to determine whether xPAPCc can activate LacZ by itself.

3.4.5. Optimization of 3-AT Concentration to Prevent His3 Leak in L40-xPAPCc

1. Streak L40-xPAPCc strain on SD-Trp/-His plates containing 0, 2.5, 5, 7.5, 10, 12.5, 15 mM 3-AT and incubate at 30°C for a week.

2. Use the lowest concentration of 3-AT that allows small colonies (<1 mm) to grow after a week.

Titer of cDNA Library

1. Take *E. coli* frozen glycerol stock (transformed with cDNA library), do not thaw it. Just scratch with a sterile needle or tip at the surface of the frozen bacteria and thaw it in a sterile Ep tube.

2. Take 2-μl thawed bacteria into 1 ml LB media and transfer to plastic cuvette.

3. Measure OD_{600}. 1 OD_{600} = 10^{6-8} viable cells per ml. Assuming an average of 10^7 cells/OD/ml, calculate the approximate concentration of viable cells in the 1-ml dilution.

4. Based on the estimated average concentration dilute 5×10^6 cells in 25 ml LB media (200 cfu/μl) (see Note 13). Mix by inverting. From that, plate 1 μl, 0.1 μl on LB/Amp plates, respectively. Incubate o/n at 37°C.

5. Count colonies on plates. With this number it is possible to calculate the titer of the cDNA library.

Amplification of Library

6. Prewarm 100 pieces of 15 cm LB/Amp plates at 37°C.

7. As before, scratch bacterial stock with a needle or tip and transfer into 1 ml LB media. Measure OD_{600}. Calculate number of viable cells based on estimated titer.

8. Dilute 5×10^6 cells in 25 ml LB media. Mix well by inverting. Plate 200 μl (4×10^4 cfu/plate) on each 15 cm LB/Amp plate. Spread bacteria with rounded glass rod. Mix the dilution always before plating the next plate. Plate a small aliquot on small plate (as above) to calculate the number of total colonies. Incubate o/n at 37°C (see Note 14).

9. Add 6 ml LB/Amp media per plate and scrape colonies as completely as possible. Add 2 ml LB/Amp media per plate to wash and collect all media together into sterile flasks (about 800 ml) (see Note 15).

10. 37°C, 200 rpm for 2 h.

11. Take out 1 ml × 8 to make glycerol stocks and freeze at −80°C.

12. Spin the bacteria by centrifugation at $6,000 \times g$ for 15 min at 4°C.

13. Freeze two thirds of the pellet at −80°C.

14. Suspend one third of the pellet in 40 ml P1 buffer (Qiagen).

15. Make 2 Maxi preps (Qiagen).

16. Yield is approximately 500 μg DNA/Maxi prep.

1. Inoculate single colony of L40-xPAPCc in 10 ml SD-Trp (in 50 ml sterile flask) and grow for 8 h at 30°C at 250 rpm.

2. Inoculate all culture from above in 100 ml SD-Trp (in 500 ml sterile flask) and grow o/n at 30°C at 250 rpm.

3. Prepare a 1:10 dilution of the o/n culture in water by diluting 100 μl culture in 900 μl water, prepare a blank by diluting 100 μl SD-Trp in 900 μl water.

4. Measure the OD_{600} of the 1:10 dilution and calculate the OD of the undiluted culture by multiplying with 10.

5. Calculate the amount of culture needed for 30 OD units. Aliquot this amount of the o/n culture into 50-ml Falcon tubes and spin down at $700 \times g$ for 5 min.

6. Resuspend the pellet in 200 ml 2× YPAD (prewarmed to 30°C) in a 1-l flask and remove 1 ml aliquot.

7. Centrifuge the 1 ml aliquot at $2,500 \times g$ for 5 min, discard the supernatant, and resuspend the pellet in water.

8. Measure the OD_{600} against a water blank, the OD_{600} should be around 0.15 (15 OD units in 100 ml = OD 0.15).

9. Grow the cells at 30°C at 250 rpm for 3–5 h to OD_{600} of 0.6 (two cell divisions).

10. Thaw 1 ml single-stranded carrier DNA, boil for 5 min, place immediately on ice. Repeat once more and carrier DNA is now ready for use.

11. Prepare the LiOAc/TE mix: 1 M LiOAc-1 ml, 10× TE pH 7.5–1 ml, ddH_2O-8 ml, total-10 ml.

12. Prepare the PEG/LiOAc master mix: 1 M LiOAc-1.5 ml, 10× TE pH 7.5–1.5 ml, 50% PEG-12 ml, total-15 ml.

13. Divide the 200 ml culture into four 50-ml Falcon tubes.

14. Centrifuge at $700 \times g$ for 5 min.

15. Resuspend each pellet in 30 ml of sterile water by vortexing.

16. Centrifuge at $700 \times g$ for 5 min.

17. Remove the supernatant, resuspend each pellet in 1 ml LiOAc/ TE mix, and transfer to Ep tube.

18. Centrifuge at $700 \times g$ for 5 min.

19. Remove the supernatant and resuspend each pellet in 600 μl of LiOAc/TE mix.

20. Set up four 50 ml Falcon tubes and add 10 μg of the cDNA library to each tube.

21. Add 200 μl carrier DNA to each tube.

22. Add 600 μl yeast cells from step 19 to each tube.

23. Vortex briefly to mix.

24. Add 2.5 ml PEG mix to each tube.

25. Vortex for 1 min to thoroughly mix all components.

26. Incubate at 30°C for 45 min at 100 rpm.

27. Add 160 μl DMSO to each tube, mix immediately by shaking.

28. Incubate at 42°C for 20 min (mix occasionally).

29. Pellet cells at $700 \times g$ for 5 min.

30. Resuspend each pellet in 3 ml 2× YPAD.

31. Let the cells recover at 30°C for 90 min at 100 rpm.

32. Pellet the cells at $700 \times g$ for 5 min.

33. Resuspend each pellet in 5 ml TE, pool into one tube (20 ml).

34. Plate 1 ml per 15 cm SD-Trp/-Leu/-His/+30 mM 3-AT plate (total 20 plates).

35. Use the remaining resuspended cells to prepare $1:10^3$ and $1:10^4$ dilutions in TE and plate on 10 cm SD-Trp/-Leu plates to calculate the transformation efficiency.

36. Seal all plates with parafilm and incubate at 30°C for 3–6 days (2–3 days for SD-Trp/-Leu plates).

37. Calculate the total number of transformants and the transformation efficiency from the number of colonies on SD-Trp/Leu plates (see Note 16).

38. Pick all big colonies that grow on SD-Trp/-Leu/-His/+30 mM 3-AT plates and restreak them on fresh SD-Trp/-Leu/-His/+30 mM 3-AT plates to make master plates.

3.4.8. X-Gal Colony-Lift Filter Assay

1. Print 50 lines in each filter and autoclave the filters. Use forceps to make each filter cling to the agar of each SD-Trp/-Leu/-His/+30 mM 3-AT plate.

2. Streak each colony from master plates on each line of the filter (50 colonies per plate) and let them grow at 30°C for 2–4 days.

3. Prepare Z buffer/X-gal solution.

4. For each plate of colonies to be assayed, presoak a sterile Whatman No. 5 or VWR grade 410 filter by placing it in 5 ml of Z buffer/X-gal solution in a clean 15 cm plate.

5. Use forceps to pick up the filter on which cells grow and immerse it completely in liquid nitrogen for 10 s.

6. Remove the filter from the liquid nitrogen and let it thaw at RT to permeabilize the cells.

7. Put the presoaked filter in a clean 15 cm plate, then carefully place the filter (colonies facing up) on the presoaked filter. Avoid trapping air bubbles under or between the filters.

8. Seal the plate and incubate the filters at 30°C and check every hour for the appearance of blue colonies. Prolonged incubation (more than 8 h) may give false positives.

3.4.9. Plasmid Recovery
from Yeast and
Retransformation
into E. coli

1. For each positive clone in X-Gal filter assay (turn blue), inoculate it in 5 ml of SD-Leu media and grow o/n at 30°C at 250 rpm till saturation.

2. Centrifuge 4,000×*g* for 5 min.

3. Resuspend in 250 µl of P1 buffer of Qiagen Miniprep Kit and add about 100 µl of 400–500 µm acid-washed glass beads (Sigma G8772).

4. Add 250 µl of P2 buffer and vortex for 5 min. Incubate for another 5 min.

5. Add 350 µl of P3 buffer and proceed as in the Kit protocol.

6. Wash the column twice with nuclease removal buffer (PB buffer) and once with wash buffer.

7. Elute the plasmid DNA with 30 µl elution buffer heated at 50°C.

8. Thaw the KC8 electrocompetent cells on ice.

9. Add 40 µl of KC8 cells to the prechilled electroporation cuvette.

10. Add 2–4 µl of eluted plasmid DNA to the cells and mix well by gently tapping the tube.

11. Perform the electroporation according to the manufacturer's instructions.

12. After shocking, immediately add 1 ml of LB media.

13. Transfer the cells to a 15-ml tube and incubate at 37°C for 1 h with vigorous shaking (250 rpm). Do not reduce incubation time.

14. Pellet cells by centrifugation at 2,500×*g* for 5 min at RT and resuspend the pellet in 100 µl of M9 media and spread on M9/Amp agar plates for nutritional selection.

15. Incubate plates at 37°C for 36–48 h and isolate plasmid from three randomly selected KC8 transformants.

16. Digest isolated prey plasmids with *Eco*RI/*Xho*I to check the size of the insert to know whether each colony contain more than one prey plasmid.

3.4.10. Confirmation
of Positive Interactions

1. Use primer pACT2-U2 and pACT2-D2 to do sequencing of each prey plasmid.

2. Do BLAST with the prey sequences.

3. Select prey plasmids encoding potential interacting proteins to confirm positive interactions.

4. Cotransform frozen L40 cells with prey plasmid and bait pNLX3–xPAPCc plasmid and plate on SD-Trp/-Leu plates to make sure that both plasmids are contransformed into L40.

5. After incubation at 30°C for 2–3 days, a lot of colonies should appear.

6. Randomly pick up 2–3 colonies from each plate and streak them as a line on SD-Trp/-Leu or SD-Trp/-Leu/-His/+15 mM 3-AT/+X-Gal plates, respectively.

7. Incubate SD-Trp/-Leu plates at 30°C for 2–3 days to make sure that the colonies grow. Incubate SD-Trp/-Leu/-His/+15 mM 3-AT/+X-Gal plates at 30°C for 4–6 days. If blue colonies grow, the interaction is interpreted as positive.

4. Notes

1. Depending on the condition of frogs, dose of 350–500 IU can be used.

2. Adjust pH to 8.0 with NaOH. Otherwise the acidic cysteine solution is deadly for the fertilized embryos.

3. Use of 2×YTA instead of YTA media help increase the expression of GST-fusion protein.

4. Incubate at 30°C might increase the yield of the full-length protein as compared to its degradation products. In some instances, temperature lower than 30°C can be applied to prevent the degradation of GST-fusion protein.

5. Sonication is not necessary if the cell pellet is treated with lysosyme.

6. Addition of Triton X-100 may aid in the solubilization of fusion proteins prior to affinity purification.

7. This helps prevent the degradation of GST-fusion protein.

8. The embryo extracts have a lot of lipids on the top, so extreme care should be taken to take the supernatant.

9. A yeast strain harboring a selectable plasmid grows much slower in appropriate selection media.

10. Too many doubling times lead to loss of selection marker.

11. Unlike *E. coli* competent cells, flash freezing of yeast in liquid nitrogen will severely reduce their competency.

12. A master mix may be made, by leaving out plasmid DNA and mixing the other three ingredients, vortex lightly to mix before aliquoting.

13. 25 ml dilution will be used later to amplify the library and is used here to have similar conditions.

14. Plate more colonies than there are independent clones in the library ($2x$–$3x$). Usually about 100 large plates are required to plate 3–4×10^6 colonies.

15. It is important to leave enough time for this step since it takes about 4 h to scrape colonies from 100 plates.

16. Total number of transformants = number of colonies on SD-Trp/Leu plate × dilution factor × 20; Transformation efficiency = total number of colonies/40 μg (clones/μg DNA). Total number of transformants should be greater than 2×10^6.

Acknowledgments

I am deeply indebted to Herbert Steinbeisser for helpful discussion and critical reading of the manuscript. I thank O. Staub for kindly providing yeast two-hybrid library and J. Moreau for pNLX3 plasmid.

References

1. Zallen JA. (2007) Planar polarity and tissue morphogenesis. *Cell* 129:1051–1063

2. Jones C, Chen P. (2008) Primary cilia in planar cell polarity regulation of the inner ear. *Curr Top Dev Biol* 85:197–224

3. Simons M, Mlodzik M. (2008) Planar cell polarity signaling: from fly development to human disease. *Annu Rev Genet* 42:517–540

4. Wang Y. (2009) Wnt/Planar cell polarity signaling: a new paradigm for cancer therapy. *Mol Cancer Ther* 8:2103–2109

5. Wang Y, Steinbeisser H. (2009) Molecular basis of morphogenesis during vertebrate gastrulation. *Cell Mol Life Sci* 66: 2263–2273

6. Medina A, Swain RK, Kuerner KM, Steinbeisser H. (2004) Xenopus paraxial protocadherin has signaling functions and is involved in tissue separation. *EMBO J* 23:3249–3258

7. Unterseher F, Hefele, JA, Giehl K, De Robertis EM, Wedlich D, Schambony A. (2004) Paraxial protocadherin coordinates cell polarity during convergent extension via Rho A and JNK. *EMBO J* 23:3259–3269

8. Wang Y, Janicki P, Köster I, Berger CD, Wenzl C, Grosshans J, Steinbeisser H. (2008) Xenopus Paraxial Protocadherin regulates morphogenesis by antagonizing Sprouty. *Genes Dev* 22:878–883

9. Debonneville C, Staub O. (2004) Participation of the ubiquitin-conjugating enzyme UBE2E3 in Nedd4-2-dependent regulation of the epithelial Na + channel. *Mol Cell Biol* 24:2397–2409

10. Iouzalen N, Camonis J, Moreau J. (1998) Identification and Characterization in Xenopus of XsmgGDS, a RalB Binding Protein. *Biochemical and Biophysical Research Communications* 250:359–363

Chapter 4

Cuticle Refraction Microscopy: A Rapid and Simple Method for Imaging *Drosophila* Wing Topography, an Alternative Readout of Wing Planar Cell Polarity

David Neff, Justin Hogan, and Simon Collier

Abstract

The polarity of hairs on the adult *Drosophila* wing provides information about the planar cell polarity (PCP) signaling events that occur during pupal wing development. We have recently shown that PCP signaling also determines the orientation of cuticle ridges that traverse the surface of the adult wing membrane; a feature we call the wing membrane topography. Although hair polarity is uniform across the wild-type wing, ridge orientation differs between the anterior and posterior wing. Consequently, mapping wing topography can provide additional information about PCP signaling, rather than simply confirming observations of wing hair polarity. Wing membrane ridges can be imaged using scanning electron microscopy, however, significant preparation time and operator expertise are required. Here, we describe cuticle refraction microscopy, a rapid and simple light microscopy method for imaging *Drosophila* wing topography.

Key words: *Drosophila*, Planar cell polarity, Wing, Hair, Membrane, Ridge, Topography

1. Introduction

The genetic analysis of epithelial planar cell polarity (PCP) originated with studies in the *Drosophila* wing almost 30 years ago (1). The wing is an excellent model tissue for studying PCP as each wing cell produces a single polarized hair that provides information about its planar polarity. In wild type flies, wing hairs point uniformly toward the distal wing margin and numerous studies have confirmed that the activities of the Frizzled PCP (Fz PCP) and Fat/Dachsous (Ft/Ds) signaling pathways are required to organize this regular array (reviewed in (2)). Recently, we have shown that the same signaling pathways determine the orientation of cuticle

Kursad Turksen (ed.), *Planar Cell Polarity: Methods and Protocols*, Methods in Molecular Biology, vol. 839,
DOI 10.1007/978-1-61779-510-7_4, © Springer Science+Business Media, LLC 2012

ridges that traverse the wing membrane, i.e., the wing topography (3–5). However, in contrast to hair polarity, ridge orientation varies between the anterior wing, where ridges run anterior to posterior, and the posterior wing, where ridges run proximal to distal (3). The different relationship between ridge orientation and hair polarity in the anterior and posterior wing implies that information gained from studying wing topography does not just confirm observations of hair polarity, but can provide additional information about wing PCP.

Wing membrane topography can be imaged using scanning electron microscopy (SEM), however, this method has at least two significant shortcomings. First, SEM requires substantial sample and microscope preparation time, and is often dependent upon securing use of a shared instrument. Second, SEM requires significant investigator training to optimize sample preparation, instrument operation, and data interpretation. In contrast, the cuticle refraction microscopy (CRM) method outlined in this chapter (see Note 1) requires minimal sample preparation time and uses a standard light microscope available to most biologists. In addition, once the optimal configuration for CRM is established for any specific light microscope (see Notes 2 and 3), little investigator training is required as successful capture of a CRM image requires only adjustment of microscope focus and illumination/exposure settings.

The basic CRM procedure is as follows: A fly wing is laid flat on a layer of mountant on a microscope slide, such that the mountant contacts the underside of the wing, but does not flow over the upper surface. The sample is sealed under a cover slip and then viewed by light microscopy with the incident light as close to collimated as possible. At a focal plane above the sample surface, membrane ridges are manifest as bright lines against a dark background (Fig. 2a). At a focal plane below the sample surface, "membrane valleys" (the troughs between the membrane ridges) are manifest as bright lines against a dark background (Fig. 2c).

The theory behind CRM is as follows: Placing a wing on fluid mountant sculpts the upper surface of the mountant into a replica of the wing topography. Since the *Drosophila* wing membrane is ridged, the surface of the mountant will be molded into ridges (see Fig. 1). Both the mountant and wing cuticle have a higher refractive index than air. Consequently, the mountant beneath a membrane ridge acts as a rudimentary plano-convex lens that converges collimated light toward a focal point above the ridge (see Fig. 1). Imaging at a focal plane above the sample surface reveals bright lines of converged light that track the tops of the wing ridges (Fig. 2a). We have previously referred to this as CRM-High (CRM-H) imaging (3). In contrast to membrane ridges, the mountant beneath a membrane valley acts as a rudimentary plano-concave lens that diverges collimated light resulting in a virtual focal point beneath the sample surface (see Fig. 1). Therefore, imaging at a

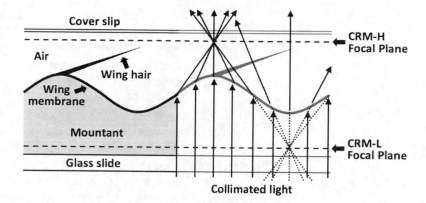

Fig. 1. Schematic of CRM method. The *left side* of the diagram shows a cross-sectional view of a *Drosophila* wing mounted for CRM analysis (Note: Individual components are not necessary to scale and some distances are exaggerated for clarity). The *black arrows* on the *right side* of the diagram indicate the path of collimated light through the sample. Note that collimated light converges above a wing membrane ridge and diverges above a wing membrane valley. The *black dashed lines* trace diverged light back toward a virtual focal point beneath a wing membrane valley.

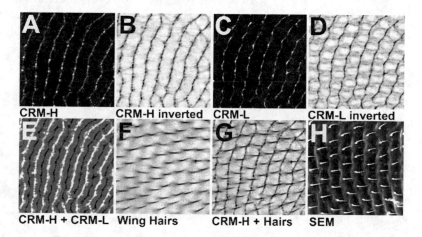

Fig. 2. Cuticle refraction microscopy (CRM). All images are of the same anterior region of the *Drosophila* wing, with anterior at top. (**a–g**) are images of the same *Drosophila* wing, (**h**) is a scanning electron micrograph of a different *Drosophila* wing. (**a**) CRM-H image. Membrane ridges (*white*) run anterior to posterior. (**b**) Inverted CRM-H image. (**C**) CRM-L image. Membrane valleys (*white*) run anterior to posterior. (**d**) Inverted CRM-L image. (**e**) Merge of CRM-H and inverted CRM-L images. Membrane ridges (*white*) map between membrane valleys (*black*). Compare this low-resolution topography map with a scanning electron micrograph of the same wing region (**h**). (**f**) Wing hair image. (**g**) Merge of inverted CRM-H image and wing hair image. Membrane ridges and wing hairs are approximately orthogonal. Compare this image with a scanning electron micrograph of the same region (**h**). (**h**) scacnning electron micrograph for comparison (note that this is not the same wing as that shown in **a–g**).

focal plane below the sample surface reveals bright lines that track the floor of wing membrane valleys (Fig. 2c). We have previously referred to this as CRM-Low (CRM-L) imaging (3). CRM-H and CRM-L images of the same wing region can be merged to generate a low-resolution map of wing membrane topography (e.g., compare Fig. 2e with Fig. 2h).

Two developments were key to the development of the CRM method. The first was to identify an appropriate mountant to support the wing. The second was to generate the approximately collimated light required for the method using a standard light microscope. An appropriate mountant needs to fully contact the underside of the wing without flowing over the top surface and should harden to make the sample permanent. Trial and error identified quick-drying clear nail polish as the optimum choice. Nail polish freshly applied to a microscope slide is fluid enough to fully contact the underside of the wing, but has sufficient surface tension to prevent substantial seepage over the upper wing surface. Ensuring that incident light is close to collimated can be achieved by minor modifications of standard light microscope although, in practice, the optimal settings for CRM will need to be established for each microscope. However, there are only a limited number of microscope modifications that can improve beam collimation (see Note 2) and the effect of these changes on collimation can easily be assayed (see Note 3 and Fig. 3).

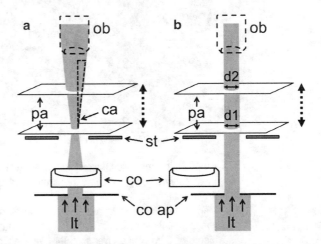

Fig. 3. Diagram of method used to assess collimation of light microscope beam (see Note 3). (**a**) Beam for normal light microscopy. (**b**) Collimated beam ideal for CRM (*ca* convergence angle, *co* condenser, *co ap* condenser aperture diaphragm, *d1* and *d2* diameter of light spot on paper, *lt* light, *ob* objective, *pa* paper, *st* stage). To test for collimation, a piece of blank paper (*pa*) is moved up and down through the microscope light beam (*dashed black arrow*). Light is best collimated when the light spot on the paper shows least variation in diameter at different positions, i.e., when diameters *d1* and *d2* are most similar and the convergence angle (*ca*) is minimal (note that the removal of the condenser is not necessarily required to achieve near collimated light).

Wings prepared for CRM can still be used to observe hair polarity (Fig. 2f), although the images generated will have more background than those from wings mounted conventionally (e.g., in GMM). This is useful, as the CRM images are particularly informative when combined with hair images to show both aspects of PCP for any specific region of the wing (Fig. 2g). However, to achieve this, CRM images need to be inverted and brightness/contrast adjusted such that the dark hairs are visible against the light background of the inverted CRM image. This CRM image processing can be undertaken using free NIH ImageJ software and is also described in this chapter.

2. Materials

2.1. Wing Mounting

1. Microscope slide: Superfrost microscope slides (3 in. × 1 in. × 1.2 mm) (Fisher, Pittsburg, PA, USA).

2. Clear Nail Polish for mounting: Super Top Speed Nail Enamel (Revlon, New York, NY, USA). Other brands of quick-drying, clear nail polish may be appropriate, but will need to be tested for effectiveness (see Note 7).

3. Cover slip: Microscope cover glass (22 mm × 22 mm, 0.16–0.19 mm thickness) (Fisher, Pittsburg, PA, USA).

4. Colored Nail Polish for sealing: Sheer Translucide (Revlon, New York, NY, USA). Most brands of quick-drying, colored nail polish can probably be used for sealing.

2.2. Microscopy

1. Light microscope: Olympus BX51 System (Olympus America Inc., Melville, NY, USA) with a 40× air objective. Most light microscopes can probably be adapted for CRM; for a general discussion of microscope settings appropriate for CRM, see Note 2.

2. Imaging Software: Olympus Microsuite Five on Microsoft Windows XP operating system. Most imaging software should be appropriate for capturing CRM images.

2.3. Image Processing

1. Image processing: NIH ImageJ software version 1.43 (http://rsb.info.nih.gov/ij/). This is free downloadable software with ongoing support and development provided by National Institutes of Health. Other image processing software (e.g., Adobe Photoshop) may be used, but are not discussed in this chapter.

3. Methods

3.1. Wing Mounting

1. Immobilize flies using CO_2 or anesthetic, and select mature, clean adult flies of the appropriate genotype (see Note 4).

2. Hold a fly steady by gripping the thorax with forceps. With a second pair of forceps, grip the wing hinge and remove the wing by pulling outward away from the thorax (see Note 5). Lay the wing on a clean dry surface (see Note 6).

3. Brush a layer of quick-drying nail polish on the surface of a microscope slide and place the wings gently on the surface of the nail polish such that mountant contacts the bottom side of each wing, but does not overflow the top surface (see Note 7). Allow slide to dry for at least 1 h.

4. Carefully place a cover slip on top of the sample, avoiding lateral movement, and seal the gap between the cover slip edges and the slide using colored nail polish (see Note 8). Allow slide to dry for at least 1 h.

5. Wings prepared for CRM may be stored indefinitely at room temperature without significant deterioration of the sample.

3.2. Microscopy

1. Place the slide on the microscope stage as for conventional light microscopy.

2. Adjust microscope settings to generate an approximately collimated light beam. For the Olympus BX51 (or similar) microscope: flip-out the condenser lens from the light path and narrow the condenser aperture diaphragm. (For a general discussion of microscope adjustments that may generate a collimated light beam, see Note 2. For a simple test for a collimated light beam, see Note 3 and Fig. 3.)

3. Adjust focal plane above the sample surface until bright lines are focused across the field of view (see Fig. 2a). The CRM-H image can then be captured.

4. Adjust focal plane below the sample surface until bright lines are focused across the field of view (see Fig. 2c). The CRM-L image can then be captured (see Note 9).

5. Without moving the slide, adjust microscope settings to those optimal for imaging wing hairs. For the Olympus BX51 (or similar) microscope: replace the condenser lens and fully open the condenser aperture diaphragm. Adjust focus to visualize wing hairs and, if needed, focus the condenser at the sample plane. A wing hair image can then be captured.

6. CRM and wing hair images should be saved in a format compatible with ImageJ. The uncompressed bitmap (.bmp) format works best in our hands.

3.3. Image Processing

1. Open ImageJ software and import CRM and wing hair images into ImageJ by dragging and dropping image files (.bmp or .tif) onto the blue ImageJ toolbar.

2. Convert images to 8-bit grayscale format to allow subsequent image merging. For each image; select *Image > type > 8-bit* and then resave image using 8-bit, .tif or .bmp format (to avoid compression artifacts, do not use the .jpg format).

3. Adjust brightness and contrast as appropriate to reduce background; select I*mage > Adjust > Brightness/Contrast.* Optimally adjusted CRM images will show almost no light intensity in the spaces between the bright lines (see Fig. 2a, c). Optimally adjusted hair images will have an almost white background (see Fig. 2f).

4. Invert contrast of CRM images (see Fig. 2b, d) to facilitate merging with other CRM or wing hair images; select *Image > Color > edit LUT > invert.* The inverted image must be saved and reopened in ImageJ to remain inverted during subsequent image merging.

5. To merge an inverted CRM image with a hair image (see Fig. 2g), first open both images. Select *Image > color > merge channels.* Select the CRM image for the red channel and the hair image for the green channel, select *None* for blue and gray channel. (Click *keep source images* to allow further adjustments if the first merge is not acceptable. Once the merge is complete, the merged color image can be converted back into 8-bit grayscale by selecting *Image > type > 8-bit.*)

4. Notes

1. We have previously called this method "cuticle refraction microscopy" or CRM (3). This name is misleading, as the light refraction required for the technique to work is largely due to the refraction differential between the mountant and air rather than between the cuticle and air. However, for consistency, we will continue to refer to this method as CRM.

2. In principle, three adjustments to a conventional light microscope can help achieve near-perfect beam collimation. The field diaphragm can be narrowed, the condenser aperture diaphragm can be narrowed, and the top lens of the condenser can be removed. (In addition, if the microscope is equipped with a high numerical aperture condenser and the top lens of the condenser is not removable, it may be beneficial to remove the entire condenser from the microscope. However, this is not ideal as wing hairs are difficult to image without a

condenser in place.) These adjustments can be made individually, or in combination, and can be done with the sample slide in place to test their effectiveness on the quality of CRM image produced. In addition, the effect of any adjustments on beam collimation can be tested using the simple procedure outlined in Note 3. In practice, we achieve near-perfect beam collimation with our Olympus BX51 microscope by flipping out the top condenser lens and closing the condenser aperture diaphragm to its narrowest. We have also achieved near-perfect beam collimation on a Nikon Diaphot300 inverted microscope simply by closing the condenser aperture diaphragm to its narrowest (this was performed with a properly Köhlered long working distance condenser with low numerical aperture; NA = 0.3).

3. Microscope beam collimation can be tested by the method diagrammed in Fig. 3. Begin by lowering the stage (*st*) or raising the objectives (*ob*) such that there is about 1 in. of free space between the stage and the objectives. Place a small piece of blank paper on the stage and note the diameter of the lighted spot (*d1*). Raise the paper approximately 1 in. from the stage surface and again note the light spot diameter (*d2*). (A microscope slide on its edge provides a convenient 1-in. riser for moving the paper from position *d1* to *d2*.) Ideally, there will be no difference between diameters *d1* and *d2* indicating that the light beam is collimated. (The convergence/divergence angle (*ca*) can be calculated by solving for the triangle shown in Fig. 3. However, this is probably not necessary as simply assessing changes in spot diameter by eye works well in our experience.) If beam divergence is noted (i.e., the spot diameter increases as the paper is raised from the stage), the microscope can be modified to try to improve beam collimation (see Note 2). The effects of these modifications on beam collimation can be tested using the same procedure.

4. Flies selected for CRM need to be old enough that all pupal wing cell debris has been cleared from the wing (6), as cell debris will block the passage of light through the wing. These flies will have a wing membrane that appears transparent, rather than translucent. Note that it is not possible to use CRM to image the topography of mutant wings in which the dorsal and ventral surfaces have separated (e.g., blistered wings) or where there is a failure of wing maturation and pupal wing cells persist between the dorsal and ventral surfaces.

5. When removing wings, it is important to avoid transferring body fluids to the slide via the forceps or the wing itself. Insect body fluids may dilute or cloud the nail polish, reducing its effectiveness as a mountant.

6. The short time available to mount wings successfully (see Note 7) means that it is advisable to lay out the wings on a

clean surface in the appropriate orientation (e.g., dorsal or ventral side uppermost), before applying the nail polish to the slide. Once the nail polish has been applied, wings can be transferred directly to the slide without losing time verifying the appropriate orientation.

7. There is a critical "window of opportunity" for placing the wing on the nail polish such that the wing lies flat on the mountant surface. If the wing is placed early, when the nail polish is too fluid, the wing margin may sink under the surface. If the wing is placed late, when the nail polish is too dry, bubbles of air may be trapped between the wing and the nail polish, or the wing may fail to adhere to the surface of the mountant. The duration of this "window of opportunity" will be dependent upon multiple factors including the brand and age of nail polish, as well as the temperature within the room. If many wings of the appropriate genotype are available, the optimal time may be determined through trial and error. However, if sample wings are limited in number, trails with wild-type wings might be undertaken initially. In practice, we find that no more than six wings may be successfully mounted for each application of nail polish.

8. When placing the cover slip on the sample, be aware that the lack of mountant above the sample means that the cover slip may directly contact parts of the wing surface. Therefore, it is important to keep any lateral movement of the cover slip at a minimum to prevent possible damage to the upper wing surface. It is also important to take care when sealing the cover slip with nail polish, as it is possible for the nail polish to seep into the air gap between the slide and cover slip and partially cover the upper wing surface. This will make the topography of this region of the wing uninterpretable. It is best to use small amounts of quick-drying, colored nail polish as it is then easy to see when the sample is completely sealed and also if any nail polish is seeping below the cover slip.

9. CRM imaging can be undertaken at high magnification (e.g., 600–1,000×), however, due to the close working distance of the objective lenses required, it is not possible to image at the virtual CRM-L focal plane (see Fig. 1).

Acknowledgments

Thanks to Kristy Doyle for her contribution to the early development of the CRM method. Thanks also to Michael Norton at the Marshall University Molecular and Biological Imaging Center (MBIC) for assistance with SEM imaging. Development of the

CRM method was supported by West Virginia NSF Experimental Program to Stimulate Competitive Research (WV-EPSCoR), the Marshall University Center for Cell Differentiation and Development (MU-CDDC) and an NSF award to Simon Collier.

References

1. Gubb, D., and Garcia-Bellido, A. (1982) A genetic analysis of the determination of cuticular polarity during development in Drosophila melanogaster, *J Embryol Exp Morphol 68*, 37–57.

2. Seifert, J. R., and Mlodzik, M. (2007) Frizzled/PCP signalling: a conserved mechanism regulating cell polarity and directed motility, *Nat Rev Genet 8*, 126–138.

3. Doyle, K., Hogan, J., Lester, M., and Collier, S. (2008) The Frizzled Planar Cell Polarity signaling pathway controls Drosophila wing topography, *Dev Biol 317*, 354–367.

4. Hogan, J., Valentine, M., Cox, C., Doyle, K., and Collier, S. (2011) Two frizzled planar cell polarity signals in the Drosophila wing are differentially organized by the fat/dachsous pathway, *PLoS Genet 7*, e1001305.

5. Valentine, M., and Collier, S. (2011) Planar cell polarity and tissue design: Shaping the Drosophila wing membrane, *Fly (Austin) 5*(4)

6. Kiger, J. A., Jr., Natzle, J. E., Kimbrell, D. A., Paddy, M. R., Kleinhesselink, K., and Green, M. M. (2007) Tissue remodeling during maturation of the Drosophila wing, *Dev Biol 301*, 178–191.

Chapter 5

Analysis of Cell Shape and Polarity During Zebrafish Gastrulation

Douglas C. Weiser and David Kimelman

Abstract

Gastrulation is a complex set of cellular rearrangements that establish the overall shape of the body plan during development. In addition to being an essential and fascinating aspect of development, the cells of the gastrulating zebrafish embryo also provide an ideal in vivo system to study the interplay of cell polarity and movement in a native 3D environment. During gastrulation, zebrafish mesodermal cells undergo a series of conversions from initial non-polarized amoeboid cell movements to more mesenchymal and finally highly polarized and intercalative cell behaviors. Many of the cellular behavior changes of these cells are under the control of the RhoA pathway, which in turn is regulated by many signals, including non-canonical Wnts. The goal of this chapter is to provide researchers with the necessary protocols to examine changes in cell polarity and movement in the developing zebrafish embryo.

Key words: Planar cell polarity, Convergent extension, Convergence and extension, Wnt, RhoA, Blebbing, Rock, Myosin, Zebrafish, Gastrulation

1. Introduction

Gastrulation is a tightly controlled set of cellular movements required to generate the vertebrate body plan (1). Convergence and extension (CE; also called convergent extension when the two processes are intimately linked) are two of the primary driving forces of gastrulation (1–4). During CE, cells migrate toward the dorsal side of the embryo and then intercalate between surrounding cells. This process results in an overall dorsoventral narrowing and anterioposterior lengthening of the embryo (1). The β-catenin-independent Wnt planar cell polarity pathway (Wnt-PCP, also called the non-canonical Wnt pathway) is a major regulator of cell polarity and cell movement in a wide array of developmental and disease systems (5–7). One context in which the Wnt-PCP has

Kursad Turksen (ed.), *Planar Cell Polarity: Methods and Protocols*, Methods in Molecular Biology, vol. 839,
DOI 10.1007/978-1-61779-510-7_5, © Springer Science+Business Media, LLC 2012

been particularly well studied is its essential role in controlling cell movements during CE (2, 3, 8, 9). Both gain- and loss-of function of members of the PCP pathway result in severe gastrulation defects in a variety of vertebrates (10–18).

One particularly powerful system for the study of CE is zebrafish. Zebrafish embryos are optically transparent during gastrulation, which makes microscopic observations using Differential Interference Contrast optics (DIC) much simpler than in many vertebrates. Another powerful advantage of zebrafish is the availability of numerous mutants, including several disrupting CE. Finally, gain- and loss-of-function experiments are quite straightforward in zebrafish, as many genes can be over-expressed by injection of mRNA or knocked-down by injection of morpholino oligonucleotides (19, 20).

During zebrafish gastrulation, mesodermal cells undergoing CE undergo a series of stereotypical cellular behavior changes. Early in gastrulation mesodermal cells have little apparent polarity yet are highly migratory using primarily amoeboid cell movement. As gastrulation proceeds, the mesodermal cells become more highly polarized and migrate using a more mesenchymal form of movement (21–23). Finally, as convergence has largely concluded, the cells reach the dorsal side of the embryo and alter their behavior, switching from primarily migration to cellular intercalation (21). These dramatic changes in cellular behavior during gastrulation mirror changes seen in metastatic cells, which can invade neighboring tissue using collective cell migration, moving by mesenchymal or amoeboid migration (24, 25).

Many genes have been identified that regulate CE (reviewed in refs. 2, 3). One common theme for many of these genes is regulation of the RhoA GTPase pathway. RhoA is one of the primary molecular targets of the Wnt-PCP pathway but many Wnt-independent pathways also regulate RhoA (26–32). However, it seems likely that many more genes involved in this complex process have not been identified. In this chapter we will provide a framework for investigators that wish to characterize the role of novel genes in gastrulation. We will describe a number of techniques to analyze zebrafish at the whole embryo, tissue, and cellular level. These protocols can be used to characterize the role of novel genes in gastrulation as well as elucidating the complex regulation of RhoA signaling during gastrulation.

2. Materials

1. Fish Water: 0.3 g/L Instant Ocean® salts in deionized water.
2. Embryo Media (EM): 15 mM NaCl, 0.5 mM KCl, 1.25 mM CaCl$_2$, 1 mM MgSO$_4$, 0.12 mM KH$_2$PO$_4$, 0.05 mM NaHPO$_4$, 0.8 mM NaHCO$_3$. A 10× solution can be stored for several

months at 4°C without the addition of the NaHCO$_3$. 1× should be made fresh each week, with freshly added bicarbonate.

3. Use either a gel-loading pipette tip (such as VWR 37001-270) or a fine hair (such as an eyebrow hair) glued in a glass capillary as a probe to orient embryos.

4. Microinjector (e.g. Picospritzer).

5. Dissecting microscope.

6. Agarose plates: Create a 1% agarose solution in 1× EM, boil in microwave and pour into a 10 cm dish creating a thin layer of agarose. Allow the agarose to solidify and cover with EM.

7. Morpholino (Danieau) buffer: 58 mM NaCl, 0.7 mM KCl, 0.4 mM MgSO$_4$, 0.6 mM Ca(NO$_3$)$_2$, 5 mM HEPES, pH 7.6.

8. Morpholinos: Stock kept at 25 mg/ml at RT (or –80°C). Dilutions in water or Morpholino buffer can be kept at room temperature for several weeks. Some investigators believe heating at 65°C for 3 min before dilution produces more consistent results.

9. 1.5, 1.75, and 2.0% methylcellulose in EM: Prewarm 1× EM to 65°C in a water bath, add appropriate amount of powdered methylcellulose and vortex. Cool the solution in an ice bucket until it becomes clear. This can be kept for 2 weeks at room temperature.

10. 1× PBS: 137 mM NaCl, 2.7 mM KCl, 10 mM Na$_2$HPO$_4$, 1.76 mM KH$_2$PO$_4$ pH 7.4.

11. 1× PBST: PBS with 0.1% Tween.

12. Pronase: 4% pronase (Protease, type XIV, Sigma P5147) in EM (store aliquots at –20°C).

13. 4% Paraformadehyde in PBS (Make fresh and store in aliquots at –20°C; thaw a fresh aliquot each week).

14. 30, 50, 70% Glycerol in PBST.

15. Antibody block: 2% sheep serum, 2 mg/ml bovine serum albumin. Store at –20°C.

16. Alkaline phosphatase buffer: 100 mM Tris–HCl, 50 mM MgCl$_2$, 100 mM NaCl, 0.1% Tween, pH 9.5. Must be made fresh.

17. Hybridization buffer: 65% formamide, 5× SSC, 500 µg/ml torula (yeast) tRNA, 50 µg/ml heparin, 0.1% Tween-20, 9 mM citric acid, pH 6–6.5.

18. 20× SSC: 3 M NaCl, 300 mM Sodium citrate dihydrate.

19. Methanol 100%.

20. DIG RNA labeling mix (Boehringer 11 277 073 910). Store at –20°C.

21. NBT Stock: 50 mg NBT (Nitrotetrazolium blue chloride; Sigma N6639) in 0.7 ml anhydrous dimethyformamide and 0.3 ml of sterile water. Store at –20°C.

22. BCIP Stock: 50 mg of BCIP (5-Bromo-4-chloro-3-indoyl phosphate disodium salt; Sigma B1026) in 1 ml of anhydrous dimethylformamide. Store at –20°C.

23. Wide tip 5¾ in. Pasteur pipettes for transferring embryos.

24. Water bath set to 65°C.

25. Incubator set to 28.5°C.

26. Depression slide.

27. In situ hybridization probes: *papc* (*paraxial protocadherin*; marker for presomitic mesoderm). Quadfecta probe: mix of *hgg1* (*hatching gland 1*), *shh* (*sonic hedgehog*), *pax2a* (*paired box gene 2a*), and *dlx3b* (*distal-less homeobox gene 3b*). Store at –20°C.

28. Glass-bottomed dishes (35 mm): MatTek Cultureware (P35G-1.5-14-C).

29. Microscope for DIC: Time-lapse recordings were performed with a Zeiss Axiovert 200M microscope, DIC optics, and a 40× objective. The experiments should be performed at constant temperature, ideally 28°C.

30. Software: AxioVision 4 software was used to collect time-lapse movies and measure angles of convergence and extension. ImageJ was used to track cell movements.

31. Caged fluorescein dextran: Currently there are no commercial suppliers of this reagent. However, a protocol for synthesizing this reagent has been developed by James Chen (Stanford) and is available at http://chen.stanford.edu/documents/Caged FluoresceinDextran.pdf.

32. Anti-digoxigenin Fab fragments conjugated to alkaline phosphatase (Roche Molecular Biochemicals). Store at 4°C.

33. Rockout: should be maintained in a stock solution in DMSO (Calbiochem 555553). Store at –20°C.

34. Blebbistatin: should be maintained in a stock solution in DMSO (Sigma B0560). Store at –20°C.

3. Methods

3.1. Injection and Dechorionation of Zebrafish Embryos

For additional protocols on the husbandry of zebrafish, embryo collection and staging, see the zebrafish book (33), which is available online at https://wiki.zfin.org/display/prot/ ZFIN+Protocol+Wiki. Morpholino antisense oligonucleotides (MOs) can be easily designed for loss-of-function experiments by either blocking splicing or translation (20) and are commercially available (www.gene-tools.com). If a gene has been hypothesized

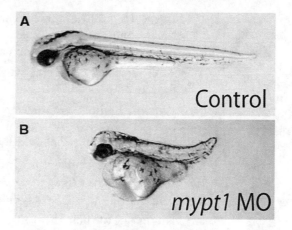

Fig. 1. CE is required for body axis elongation in zebrafish. (a) Dorsal views of 48 hpf zebrafish embryos injected with (a) morpholino buffer or (b) 1 ng *mypt1* MO. Knock-down of Mypt1 causes dramatic reduction in both convergence and extension (26), leading to a dramatically shortened and curved body axis.

to play a role in gastrulation it can be knocked down by injection of a MO into the one-cell stage embryo. Although we describe here loss-of-function approaches, the same methods can be used for gain-of-function experiments using synthetic mRNA.

The most obvious phenotype of an embryo with severe defects in CE is a shortened and curved body axis at 48 hpf (hours post-fertilization). Figure 1 shows a 48 hpf embryo injected with buffer control or an MO targeting *mypt1*, a gene required for both convergence and extension (26). Loss of some regulators of CE have a dramatically shortened body axis as shown in Fig. 1 (15), whereas others have only mild body axis curvature and shortening (10). Because MOs can have off-target effects (20) that can resemble CE defects, it is important either to show that the MO effect can be rescued by co-injection of a synthetic mRNA for the gene of interest that will not bind to the MO or to show that two different MOs cause the same effect. It is often useful to test a MO with a 5-bp mismatch to ensure that the phenotype is specific (we do not use the standard control MO sold by Gene Tools since it seems to be chosen precisely because it has no effects).

1. Allow zebrafish to spawn naturally and collect embryos within 20 min of spawning.

2. Place 50–100 embryos in an injection tray with a minimal amount of water.

3. Place the desired buffer to be injected into an injection needle. Calibrate the injection time with a reticule to deliver 1 nl of buffer per injection.

4. Inject 100–300 embryos with the MO (diluted in water or Morpholino buffer). Typical MO range is 1–20 ng, although doses above 5 ng can often cause non-specific effects.

5. After injecting place embryos in petri dish with EM and incubate at 28.5°C until the desired stage is reached.

6. When the embryos approach 50% epiboly prepare to dechorionate them. They can be manually dechorionated with fine forceps or using pronase as in the following steps.

7. Place 50–100 embryos in 1 ml EM in a small petri dish covered with a thin layer of agarose. Add 1 ml of 4% pronase in EM.

8. Swirl gently and watch for 2–4 min for the first sign of embryos falling out of their chorions.

9. Pour the embryos into 300 ml of EM in a 500 ml or 1 L beaker with a thin layer of agarose at the bottom.

10. Allow embryos to fall to the bottom and pour off most of the liquid.

11. Add 300 ml of fresh EM gently and pour off the excess.

12. Repeat once more.

13. Embryos should now be entirely or mostly dechorionated (see Note 1). Remove with a Pasteur pipette and place in another 10 cm plate with an agarose bottom.

3.2. In Situ Probe Synthesis

Note: all water and solutions used from step 2 onward must be nuclease free, either purchased as such or diethyl pyrocarbonate-treated.

1. Digest 10 μg of DNA with the appropriate restriction enzyme. Extract the aqueous solution twice with phenol-chloroform.

2. Ethanol precipitate the linearized DNA and resuspend in 10 μl of water.

3. Place 1 μg of linearized DNA, 1 μl RNAse inhibitor, 2 μl of 10× transcription buffer, 2 μl of 10× DIG RNA labeling mix, 2 μl of the appropriate polymerase (T7, T3, or SP6), bring to a total of 20 μl with water. Incubate the reaction mix at 37°C for 2 h.

4. Add 1 μl RNase-free DNase and incubate at 37°C for 30 min.

5. Add 25 μl of 7.5 M LiCl. Incubate at –20°C for at least 30 min.

6. Spin the solution in a microcentrifuge at maximum speed for 15 min at 4°C.

7. Wash the pellet with 70% ethanol. Remove all ethanol and air-dry pellet for 5 min.

8. Resuspend the pellet in 30 μl nuclease-free water and add 170 μl of hybridization buffer.

3.3. Phenotypic Characterization Using In Situ Hybridization

After examining the phenotype it is essential to carefully characterize the effect at the tissue level. In situ hybridization provides an invaluable tool to measure the extent of CE in developing embryos. Staining bud-stage embryos (end of gastrulation) allows for the direct measurement of the length and width of the paraxial mesoderm using a *paraxial protocadherin* (*papc*) probe. Figure 2 shows a control embryo and one defective in CE. Note that the paraxial mesoderm in the CE-defective embryo is dramatically wider and shorter (Fig. 2e, f). In addition, both the movement of the prechordal plate and the morphogenesis of the notochord can be monitored in the same experiment. This can be accomplished by creating a mix called the "quadfecta" of *hgg1* (*hatching gland 1*- to mark the prechordal plate), *shh* (*sonic hedgehog*- to mark the midline), *pax2a* (*paired box gene 2a*- to mark the midbrain–hindbrain boundary), and *dlx3b* (*distal-less homeobox gene 3b*- to mark the neural plate). Note the dramatically broader notochord indicating defective CE of

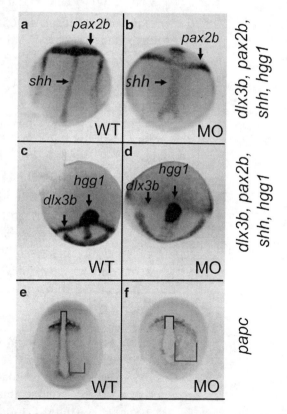

Fig. 2. Disruption of CE leads to characteristic morphological defects in zebrafish embryos. Embryos at bud stage stained with (**a–d**) *hgg1* (to mark the prechordal plate), *shh* (midline), *pax2a* (midbrain–hindbrain boundary), and *dlx3b* (neural plate) or (**e**, **f**) *papc* (presomitic mesoderm). The *narrow bracket* marks the width of the notochord and the *wide bracket* marks the width of the presomitic mesoderm. *MO* embryos were injected with 1 ng *mypt1* MO.

the axial mesoderm (Fig. 2c, d) with the quadfecta stain. Also, note that the prechordal plate has failed to migrate beyond the midbrain–hindbrain boundary (Fig. 2a, b).

1. Collect injected and control embryos at bud stage and place in 1.5 ml microfuge tube. Fix by placing (25–100) embryos in 4% PFA either for 2 h at room temperature or overnight at 4°C.

2. Wash the embryos three times with 1 ml PBS to remove the PFA and then add 1 ml of 100% methanol (MeOH). These embryos will now dehydrate and can be stored for years at –20°C.

3. To start experiment rehydrate stored embryos by washing for 5 min with each of the following: 75% MeOH/25% PBS, 50% MeOH/50% PBS, 25% MeOH/75% PBS, and PBST.

4. Add 0.5 ml of hybridization buffer (HB) and incubate for 1–5 h at 65°C.

5. Remove hybridization mix and add 0.2 ml of desired probe (diluted 1:200 in HB) and incubate overnight at 65°C.

6. Preheat wash solutions to 65°C in a water bath. Remove the probe mix and return to –20°C. Perform the following washes at 65°C by placing the tubes in the water bath during washes. Wash for 1 min in HB, 15 min in 75% HB/25% 2× SSC, 15 min in 50% HB/50% 2× SSC, 15 min 25% HB/75% 2× SSC, twice for 10 min in 2× SSC, and twice for 30 min in 0.2× SSC. The following washes are done at room temperature for 5 min each: 75% 0.2× SSC/25% PBST, 50% 0.2× SSC/50% PBST, 25% 0.2× SSC/75% PBST, and PBST.

7. Add 0.5 ml of antibody block solution and incubate for 1–3 h at room temperature, without agitation.

8. Remove the block and add 200–500 μl antibody solution (1:5,000 anti-digoxigenin Fab fragments conjugated to alkaline phosphatase in block) then incubate overnight at 4°C.

9. Wash the embryos six times with PBST for 15 min each and three times for 5 min with alkaline phosphatase buffer (without BCIP/NBT).

10. Prepare a stock of 5 ml of alkaline phosphatase buffer and add 22.5 μl of NBT and 17.5 μl of BCIP. Add 0.5 ml of the alkaline phosphatase buffer with BCIP/NBT to the embryos. Allow the embryos to incubate at room temperature, protected from direct light until strong blue staining is visible but little background has developed (around 1 h with these probes, but varies and may need to be optimized by eye for each experiment).

11. Stop the reaction by adding 1 ml of PBS, removing the PBS and adding 1 ml of methanol. The embryos can be stored at –20°C for long periods. The methanol step is not necessary but it clears the yolk, which helps imaging the stain.

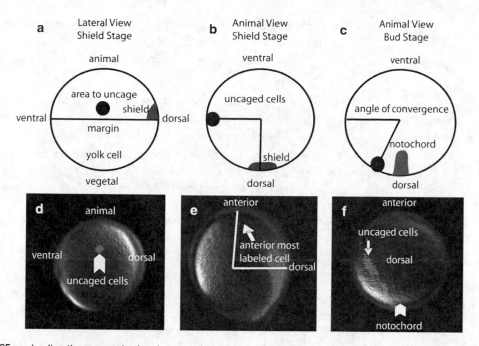

Fig. 3. CE can be directly measured using dye uncaging. (**a**) A diagram demonstrating the correct lateral orientation for dye uncaging. A shield stage embryo with the margin, shield (*gray*) and the desired area to uncage (*black dot*) are all labeled. (**b**) A diagram of an animal view of an embryo immediately after uncaging. A cluster of cells 90° from dorsal have been uncaged. (**c**) A diagram of an animal view of a bud-stage embryo after the labeled cells have undergone CE. The labeled cells have migrated toward the dorsal side. The angle between their original lateral position and their final position is the angle of convergence. (**d**–**f**) The fluorescent dye was uncaged at shield stage (**d**, *white arrow*) and examined at bud stage in lateral (**e**) and dorsal (**f**) views with the animal pole at *top*. In (**f**) the *white arrowheads* indicate the position of the notochord. (**e**) Anterior extension is defined as the angle between the anterior-most labeled cell and the midpoint along the anterior–posterior axis on the dorsal side (lateral view, anterior at *top*).

12. To continue, rehydrate the embryos (as in step 3). Wash once with 30% glycerol in PBS, once with 50% glycerol in PBS and once with 70% glycerol in PBS. Allow each wash to proceed until the embryos sink to the bottom of the tube.

13. Place embryos in depression slide and orient with a gel-loading pipette tip and image with a dissecting microscope and digital camera.

3.4. Tracking Cell Movement (Dye Uncaging)

The extent of both convergence and extension can be directly measured in zebrafish embryos (34). Embryos are loaded with caged dye and grown to the onset of gastrulation, shield stage. At this time the dye is uncaged by a brief pulse of UV radiation to a small area of the embryo. This will causes a small cluster of cells to fluoresce. An initial starting image is then collected, and the embryo is allowed to proceed through gastrulation. At bud stage the embryos are reimaged and the extent of dorsal (convergence) and anterior (extension) migration can be determined. Figure 3 shows the migration of lateral cells in a control embryo.

1. Inject at the one cell stage with 1 nl of 0.5% caged fluorescein dextran (with or without the MO).

2. Grow the embryos at 28.5°C and dechorionate (see Subheading 3.1).

3. Mount a single embryo in a drop of 1.75% methylcellulose in embryo medium at shield stage in a glass-bottomed dish (see Notes 2 and 3).

4. Carefully orient the embryos with a gel loading tip such that the shield is to the side and the lateral margin faces directly down in a glass-bottomed dish (Fig. 3a shows the way the embryo should look viewing it with an inverted microscope). Only work with 10–20 embryos at a time, each in their own dish.

5. Gently place the dish on the inverted fluorescent microscope. Find the desired cells using a 5 or 10× objective and then switch to 40×. Close the fluorescence pinhole iris to the narrowest setting. This will allow a narrow beam of light to illuminate only a small portion of the field.

6. Open the DAPI channel for approximately 15 s. This UV radiation will uncage the dye only in the portion of the embryo that is illuminated. In order to measure convergence and extension of the lateral mesoderm uncage a region of cells 90° from the dorsal shield and just above the margin (Fig. 3a, d).

7. Turn off the DAPI channel and zoom out to 5× and open the pinhole (see Note 4).

8. Take a brightfield and a fluorescent image of the embryo showing that a few cells are now fluorescing green (Fig. 3d). Acquire images with both a lateral view and animal view of uncaged embryos. This will provide you with an image of the embryo before CE and can be used to confirm that the labeled cells start 90° from dorsal.

9. Place a lid on the dish (important to prevent drying out) and place in air incubator at 28.5°C.

10. Wait approximately 4 h until gastrulation is completed and the embryos reach bud stage.

11. Photograph at 5× and acquire a dorsal view (with the notochord directly facing down in the dish), a lateral view (with the notochord oriented perpindicular to the bottom of the dish), and an animal view (with the animal pole facing directly down; Fig. 3e, f shows lateral and dorsal views, and Fig. 3c shows a diagram of an animal pole view).

12. The lateral view will provide a clear picture of the extent of extension of the embryo. Measure the angle (angles can be easily measured using either Axiovision software or ImageJ) from dorsal to the anterior most fluorescent cells (Fig. 3e). This is the angle of extension of the paraxial mesoderm.

13. The animal pole view can be easily used to measure the extent of dorsal convergence. Measure the angle from the middle of the fluorescent cells to dorsal. Immediately after uncaging this angle should be 90°. At bud stage the angle will have been reduced by the migration of the labeled cells. This difference is the angle of convergence of the paraxial mesoderm (Fig. 3b, c).

14. The dorsal view is valuable to determine how well the labeled cells have migrated dorsally and approached the notochord (Fig. 3f). This angle is, however, difficult to use to determine either convergence or extension (see Note 5).

3.5. Length–Width Ratios of Mesodermal Cells

1. Length–width ratios of cells can be easily determined from DIC images (see Note 6) and are a useful assay for cell polarity of cells undergoing gastrulation, especially since disruption of planar-cell polarity signaling can reduce the length–width ratio considerably (13, 15, 36). Cell shape can be examined at a variety of stages starting at bud stage and continuing through somitogenesis. It is important to compare identical stages when comparing experimentally manipulated and control embryos, because cell shape will change considerably and length–width ratios will increase as development proceeds.

2. Observe a field of cells that contains a clear image of the notochord and the paraxial mesoderm. Collect clear images of 10–25 notochordal cells and 10–25 mesodermal cells at least one-cell diameter away from the notochord to analyze. Collect images from at least three different embryos.

3. Measure the length of the cells using the measure function (available in either imageJ or Axiovision software) using the longest part of the cell. Then calculate the width of the cell by measuring the distance across the midpoint of the cell (Fig. 4).

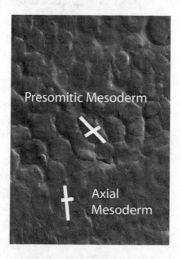

Fig. 4. Mesodermal cell polarity. The cell polarity of presomitic and notochordal mesodermal cells was determined by calculating the length–width ratio. Wild-type mesodermal cells with long axis length and short axis width are shown.

4. The length–width ratio of paraxial mesodermal cells at bud stage should be approximately 2:1, whereas notochordal cells should have a length–width ratio of approximately 3:1. The notochordal ratio will continue to increase during somitogenesis, reaching 6:1 by the 10-somite stage.

5. If desired, the length–width ratios can be calculated for ectodermal cell using the same procedure. This can be used to determine if the gene of interest is affecting only mesoderm (32) or multiple germ layers (35).

3.6. Analyzing Cell Shape and Movement

A great strength of the zebrafish system is the ability to directly watch the cell behaviors and track cell movements in real time. Investigators are not limited to only observing the phenotype after gastrulation has finished. Zebrafish mesodermal cells undergo a variety of cellular behaviors as they undergo CE including non-polarized amoeboid movement and more polarized migratory and intercalative behaviors. Mesodermal cells are capable of a several types of membrane protrusions during gastrulation, including blebs, lamellipodia, and filopodia (26, 32). During normal gastrulation, protrusive activity is dramatically reduced late in gastrulation as cells switch from migratory to intercalative behaviors. This protocol will give investigators the ability to create time-lapse images, which will allow them to observe these changes in cellular behavior.

1. Mount 1–5 dechorionated embryos in a drop of 1.5% methylcellulose in EM in a glass-bottomed dish.

2. Carefully orient the embryos with a gel-loading pipette tip under a dissecting microscope such that dorsal is facing directly downward and both the notochord and paraxial mesoderm are clearly visible.

3. Gently place embryos on the stage of a microscope with DIC optics.

4. Find the desired cells using the 5 or 10× objective. Switch to the 40× objective and focus through the enveloping layer cells, which have an epithelial shape, until you reach the more mesenchymal mesodermal cells. It is important to be consistent with the stage, location and layer of the cells to be imaged. The notochord provides a valuable landmark at bud stage, whereas the margin and shield are useful at earlier stages.

5. Once you have located the desired cells prepare a time-lapse recording. In order to track cell migration, imaging every 30 s to 1 min for 30 min is sufficient and will help reduce file size. In order to assay cell protrusive activity a faster frame rate is required, usually 3 s frames for 15 min is sufficient.

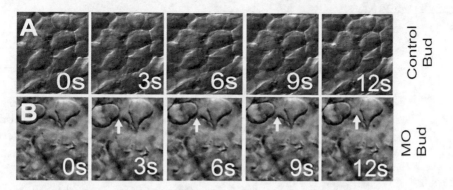

Fig. 5. Mesodermal cell shape changes dramatically during zebrafish gastrulation. Membrane-protrusive activity of mesodermal cells at bud stage was determined by time-lapse microscopy. Behavior of dorsal mesodermal cells in control embryos (**a**) and *mypt1* morphants (**b**) are shown at bud stage. *White arrows* indicate bleb-like protrusions (*s* seconds).

6. The time-lapse movie will be quite large and will need to be compressed for further manipulation. In order to quantify membrane protrusions chose 20–40 mesodermal cells per embryo from at least three embryos. Simply follow the cells through the course of the movie and score them for the number of membrane protrusions. DIC optics provides excellent resolution of blebs, lamellipodia, and filopodia in these cells (Fig. 5).

3.7. Pharmacological Inhibition of Rho-Rock Signaling

In many systems, including gastrulation, one of the key pathways that regulates membrane blebbing is the Rho-Rock-myosin pathway. Commercially available pharmacological inhibitors of Rock and Myosin are available to block membrane blebbing.

1. Remove dechorionated embryos from the incubator.

2. Prepare a 48-well culture plate by placing a few drops of melted 1% agarose in EM in the bottom of several wells.

3. Prepare 0.5–1 ml treatment solutions of 50 µM myosin inhibitor Blebbistatin and 200 µM Rockout (Rho kinase inhibitor III) or a DMSO control in EM in microfuge tubes.

4. Pipette embryos with a minimal amount of water directly into the microfuge tubes containing 1 ml of the drug in embryo media. Then transfer the embryos and media to an agarose coated well in a 48-well plate (see Notes 7–9).

5. Both drugs will affect cell shape and behavior within approximately 20 min.

6. After 30 min embryos are transferred to methylcellulose (with the same concentration of drug) and placed in a glass-bottomed dish and imaged as in protocol in Subheading 3.6.

4. Notes

1. Dechorionated embryos are extremely fragile and care should be taken in moving them, especially in methylcellulose. Embryos before bud stage should never be exposed to air since this will cause the yolk cell to burst. To orient the embryos use a single gel-loading pipette tip, or a fine hair probe to orient the embryos.

2. Embryos left for prolonged time in methylcellulose need to be treated with care. It is essential that the methylcellulose not be allowed to dry out. This will cause the embryo to burst. It is best to leave a lid on the dish with the embryos at all times when not filming.

3. The caged fluoroscein dextran is not well tolerated by embryos. Great care should be taken to not add too much dye to the embryos; however, using too little dye will result in a weak signal. Unhealthy embryos will produce auto-fluorescence of the dye, and can be used as an indicator of injecting too high a concentration. Dye injection will also produce a noticeable developmental delay and great care should be taken to carefully control all experiments.

4. The dye can be uncaged by prolonged exposure to visible light and the embryos should be protected from visible light as much as possible. Only a brief exposure to UV light is necessary to uncage the dye, and prolonged exposure can damage the embryo.

5. The dye uncaging protocol can also be used to measure extension of the axial mesoderm. The protocol is the same as Subheading 3.4 with the exception that the dorsal shield is uncaged instead of lateral cells. At bud stage the angle between the anterior most labeled cell and the midpoint along the anterior–posterior axis on the dorsal side is measured; however, unlike in protocol Subheading 3.4, these labeled cells will be in the notochord instead of the paraxial mesoderm.

6. Best results for time-lapse microscopy are obtained using DIC optics but the cells are sufficiently easy to image that all experiments can be done with bright-field optics.

7. Rockout will often precipitate if added directly to an aqueous solution without sufficient mixing. For this reason we often place the stock solution in a new tube first and then vigorously add the buffer when making a dilution.

8. The 48-well culture dishes can be washed and reused as it is not essential that they be sterile.

9. Although Y27632 is widely used to inhibit Rock in a variety of systems, it does not work as well as Rockout in zebrafish.

Acknowledgments

DK is supported by NIH grant HD27262. DCW is supported by University of the Pacific start-up funds.

References

1. Keller, R. (2002). Shaping the vertebrate body plan by polarized embryonic cell movements. *Science*, **298**, 1950–1954.

2. Hammerschmidt, M. and D. Wedlich. (2008). Regulated adhesion as a driving force of gastrulation movements. *Development*, **135**, 3625–3641.

3. Solnica-Krezel, L. (2006). Gastrulation in zebrafish – all just about adhesion? *Curr Opin Genet Dev*, **16**, 433–441.

4. Wallingford, J.B., S.E. Fraser, and R.M. Harland. (2002). Convergent extension: the molecular control of polarized cell movement during embryonic development. *Dev Cell*, **2**, 695–706.

5. Mlodzik, M. (2002). Planar cell polarization: do the same mechanisms regulate Drosophila tissue polarity and vertebrate gastrulation? *Trends Genet*, **18**, 564–571.

6. Veeman, M.T., J.D. Axelrod, and R.T. Moon. (2003). A second canon. Functions and mechanisms of beta-catenin-independent Wnt signaling. *Dev Cell*, **5**, 367–377.

7. Fanto, M. and H. McNeill. (2004). Planar polarity from flies to vertebrates. *J Cell Sci*, **117**, 527–533.

8. Kodjabachian, L., I.B. Dawid, and R. Toyama. (1999). Gastrulation in zebrafish: what mutants teach us. *Dev Biol*, **213**, 231–245.

9. Roszko, I., A. Sawada, and L. Solnica-Krezel. (2009). Regulation of convergence and extension movements during vertebrate gastrulation by the Wnt/PCP pathway. *Semin Cell Dev Biol*, **20**, 986–997.

10. Veeman, M.T., D.C. Slusarski, A. Kaykas, S.H. Louie, and R.T. Moon. (2003). Zebrafish prickle, a modulator of noncanonical Wnt/Fz signaling, regulates gastrulation movements. *Curr Biol*, **13**, 680–685.

11. Ohkawara, B., T.S. Yamamoto, M. Tada, and N. Ueno. (2003). Role of glypican 4 in the regulation of convergent extension movements during gastrulation in Xenopus laevis. *Development*, **130**, 2129–2138.

12. Carreira-Barbosa, F., M.L. Concha, M. Takeuchi, N. Ueno, S.W. Wilson, and M. Tada. (2003). Prickle 1 regulates cell movements during gastrulation and neuronal migration in zebrafish. *Development*, **130**, 4037–4046.

13. Jessen, J.R., J. Topczewski, S. Bingham, D.S. Sepich, F. Marlow, A. Chandrasekhar, and L. Solnica-Krezel. (2002). Zebrafish trilobite identifies new roles for Strabismus in gastrulation and neuronal movements. *Nat Cell Biol*, **4**, 610–615.

14. Winklbauer, R., A. Medina, R.K. Swain, and H. Steinbeisser. (2001). Frizzled-7 signalling controls tissue separation during Xenopus gastrulation. *Nature*, **413**, 856–860.

15. Topczewski, J., D.S. Sepich, D.C. Myers, C. Walker, A. Amores, Z. Lele, M. Hammerschmidt, J. Postlethwait, and L. Solnica-Krezel. (2001). The zebrafish glypican knypek controls cell polarity during gastrulation movements of convergent extension. *Dev Cell*, **1**, 251–264.

16. Wallingford, J.B., B.A. Rowning, K.M. Vogeli, U. Rothbacher, S.E. Fraser, and R.M. Harland. (2000). Dishevelled controls cell polarity during Xenopus gastrulation. *Nature*, **405**, 81–85.

17. Heisenberg, C.P., M. Tada, G.J. Rauch, L. Saude, M.L. Concha, R. Geisler, D.L. Stemple, J.C. Smith, and S.W. Wilson. (2000). Silberblick/Wnt11 mediates convergent extension movements during zebrafish gastrulation. *Nature*, **405**, 76–81.

18. Djiane, A., J. Riou, M. Umbhauer, J. Boucaut, and D. Shi. (2000). Role of frizzled 7 in the regulation of convergent extension movements during gastrulation in Xenopus laevis. *Development*, **127**, 3091–3100.

19. Nasevicius, A. and S.C. Ekker. (2000). Effective targeted gene 'knockdown' in zebrafish. *Nat Genet*, **26**, 216–220.

20. Eisen, J.S. and J.C. Smith. (2008). Controlling morpholino experiments: don't stop making antisense. *Development*, **135**, 1735–1743.

21. Yin, C., M. Kiskowski, P.A. Pouille, E. Farge, and L. Solnica-Krezel. (2008). Cooperation of polarized cell intercalations drives convergence and extension of presomitic mesoderm during zebrafish gastrulation. *J Cell Biol*, **180**, 221–232.

22. Concha, M.L. and R.J. Adams. (1998). Oriented cell divisions and cellular morphogenesis in the zebrafish gastrula and neurula: a time-lapse analysis. *Development*, **125**, 983–994.

23. Sepich, D.S., C. Calmelet, M. Kiskowski, and L. Solnica-Krezel. (2005). Initiation of convergence and extension movements of lateral

mesoderm during zebrafish gastrulation. *Dev Dyn*, **234**, 279–292.

24. Charras, G. and E. Paluch. (2008). Blebs lead the way: how to migrate without lamellipodia. *Nat Rev Mol Cell Biol*, **9**, 730–736.

25. Wolf, K. and P. Friedl. (2006). Molecular mechanisms of cancer cell invasion and plasticity. *Br J Dermatol*, **154**, 11–15.

26. Weiser, D.C., R.H. Row, and D. Kimelman. (2009). Rho-regulated myosin phosphatase establishes the level of protrusive activity required for cell movements during zebrafish gastrulation. *Development*, **136**, 2375–2384.

27. Zhu, S., L. Liu, V. Korzh, Z. Gong, and B.C. Low. (2006). RhoA acts downstream of Wnt5 and Wnt11 to regulate convergence and extension movements by involving effectors Rho kinase and Diaphanous: use of zebrafish as an in vivo model for GTPase signaling. *Cell Signal*, **18**, 359–372.

28. Habas, R. and X. He. (2006). Activation of Rho and Rac by Wnt/frizzled signaling. *Methods Enzymol*, **406**, 500–511.

29. Kwan, K.M. and M.W. Kirschner. (2005). A microtubule-binding Rho-GEF controls cell morphology during convergent extension of Xenopus laevis. *Development*, **132**, 4599–4610.

30. Jopling, C. and J. den Hertog. (2005). Fyn/Yes and non-canonical Wnt signalling converge on RhoA in vertebrate gastrulation cell movements. *EMBO Rep*, **6**, 426–431.

31. Tahinci, E. and K. Symes. (2003). Distinct functions of Rho and Rac are required for convergent extension during Xenopus gastrulation. *Dev Biol*, **259**, 318–335.

32. Weiser, D.C., U.J. Pyati, and D. Kimelman. (2007). Gravin regulates mesodermal cell behavior changes required for axis elongation during zebrafish gastrulation. *Genes Dev*, **21**, 1559–1571.

33. Westerfield, M. (1993) The Zebrafish book: a guide for the laboratory use of zebrafish (Brachydanio rerio), Eugene, OR: University of Oregon Press.

34. Sepich, D.S. and L. Solnica-Krezel. (2005). Analysis of cell movements in zebrafish embryos: morphometrics and measuring movement of labeled cell populations in vivo. *Methods Mol Biol*, **294**, 211–233.

35. Marlow, F., J. Topczewski, D. Sepich, and L. Solnica-Krezel. (2002). Zebrafish Rho kinase 2 acts downstream of Wnt11 to mediate cell polarity and effective convergence and extension movements. *Curr Biol*, **12**, 876–884.

Chapter 6

Analyzing Planar Cell Polarity During Zebrafish Gastrulation

Jason R. Jessen

Abstract

Planar cell polarity was first described in invertebrates over 20 years ago and is defined as the polarity of cells (and cell structures) within the plane of a tissue ,such as an epithelium. Studies in the last 10 years have identified critical roles for vertebrate homologs of these planar cell polarity proteins during gastrulation cell movements. In zebrafish, the terms convergence and extension are used to describe the collection of morphogenetic movements and cell behaviors that contribute to narrowing and elongation of the embryonic body plan. Disruption of planar cell polarity gene function causes profound defects in convergence and extension creating an embryo that has a shortened anterior–posterior axis and is broadened mediolaterally. The zebrafish gastrula-stage embryo is transparent and amenable to live imaging using both Nomarski/differential interference contrast and fluorescence microscopy. This chapter describes methods to analyze convergence and extension movements at the cellular level and thereby connect embryonic phenotypes with underlying planar cell polarity defects in migrating cells.

Key words: Convergence and extension, Ectoderm, Gastrulation, Nomarski/DIC, Mesoderm, Planar cell polarity, Van gogh-like 2, Wnt, Zebrafish

1. Introduction

In the fruit fly, *Drosophila melanogaster*, planar cell polarity (PCP) occurs in cuticular structures including the wing, abdomen, and eye. In the wing epithelium, PCP is manifested as the production by each cell of a single actin-rich hair that is oriented distally within the plane of the tissue (1). Mutations in PCP genes including *frizzled*, *dishevelled*, and *Van Gogh* produce stereotypic mutant polarity patterns characterized by defects in the number of hairs per cell and the subcellular prehair initiation site (2). Large-scale zebrafish mutagenesis screens identified several mutant lines with defects in gastrulation convergence and extension cell movements (3, 4).

Kursad Turksen (ed.), *Planar Cell Polarity: Methods and Protocols*, Methods in Molecular Biology, vol. 839,
DOI 10.1007/978-1-61779-510-7_6, © Springer Science+Business Media, LLC 2012

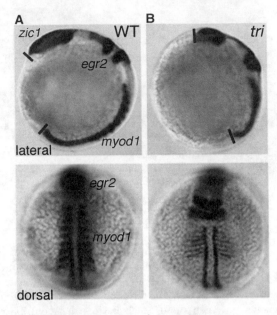

Fig. 1. *trilobite* (*tri/vangl2*) mutant embryos have a strong convergence and extension defect. (**a**, **b**) Lateral and dorsal views of wild-type and homozygous *tri^{m209}* mutant embryos at the 8–9 somite stage. Whole-mount in situ hybridization was used to highlight specific tissues including anterior neuroectoderm (*zic1*, formally *opl*), hindbrain (*egr2*, formally *krox20*), and axial mesoderm (*myod1*).

Notably, *trilobite* (*tri*) mutant embryos were shown to have mutations in *van gogh-like 2* (*vangl2*), a homolog of fly *Van Gogh* (5). Loss of Vangl2 function produces a typical morphological phenotype indicative of disrupted convergence and extension (6) (Fig. 1). This phenotype can be visualized under a simple stereomicroscope using either live embryos or fixed embryos probed with tissue-specific markers using whole-mount in situ hybridization.

A key question that emerged after the discovery that zebrafish homologs of fly PCP genes were required for morphogenetic events during gastrulation was elucidating the nature of PCP in migrating gastrula cells (7). In wild-type embryos, it was shown that mesodermal and ectodermal gastrula cells become morphologically polarized (5, 8–12). Here, polarization is defined as both an increase in a cells length–width ratio (LWR) and a change in cellular orientation such that it is parallel to the path of migration or oriented cell division. Significantly, this polarity is lost in *tri/vangl2* mutant embryos and in other embryos where components of the PCP signaling pathway have been disrupted (5, 12). During the last several years, numerous genes have been linked to convergence and extension based on morphological phenotypes assessed by in situ hybridization and by more advanced methods such as photo-activated fluorescent labeling of cell populations (13). By contrast, few proteins required for convergence and extension have been analyzed at the cellular level and thus it is unknown whether they directly impact PCP. Such data is critical if we are to understand how PCP is established and how it is coordinated with other processes regulating cell migration during gastrulation. Here, I describe methods for quantifying the PCP of individual mesodermal and ectodermal cells using agarose-mounted live embryos and Nomarski/differential interference contrast (DIC) microscopy. These methods also provide a foundation for more advanced quantitative assays of PCP using fluorescently labeled cells and confocal microscopy.

2. Materials

1. Wild type, mutant, or mRNA/morpholino-injected embryos raised at 28.5°C and 32°C (see Note 1) until the end of gastrulation (tailbud stage, 10 h post-fertilization when raised at 28.5°C) (14).

2. Egg water: 60 mg/ml Instant Ocean sea salt in ultrapure water.

3. Dry incubators set at 28.5 and 32°C (see Note 1).

4. 100% Danieau's buffer: 58 mM NaCl, 0.7 mM KCl, 0.4 mM $MgSO_4$, 0.6 mM $Ca(NO_3)_2$, 5 mM HEPES, pH 7.6.

5. Dumont Biologie tip #55 fine forceps (Fine Science Tools, cat. no. 11255-20).

6. Glass petri dishes for dechorionating live embryos.

7. 0.16–0.19 mm glass coverslip dishes (In Vitro Scientific, cat. no. D29-14-1.5 N or MatTek, cat. no. P35GC-1.5-14-C).

8. 3% Agarose in 30% Danieau's buffer.

9. 0.8% Low-melting point (LMP) agarose in 30% Danieau's buffer.

10. Microwave oven.

11. Transfer pipets (Thermo Fisher Scientific, cat. no. 13-711-7), glass Pasteur pipets, and a pipet pump dispenser.

12. Heat block set at 42°C.

13. Dissecting/stereomicroscope. We use a Leica MZ7.5 with a tilting mirror base that allows adjustable brightfield and oblique illumination techniques.

14. Motorized (including Z-axis) inverted compound microscope equipped with Nomarski/DIC optics and a 40× air or 40× oil immersion objective. Low power objectives (e.g., 10× or 20×) are also useful for documenting embryo orientation. We use a Leica DMI6000B inverted microscope equipped with 40× air (Leica Plan Apo, NA=0.85) and 40× oil immersion (Leica

Plan Apo, NA = 1.25) objectives. This microscope also has a heated stage insert for 35 mm petri dishes (Binomic cell and controller, 20/20 Technologies, Inc.). Our microscope is connected to a Dell computer workstation and we use an ORCA ER digital camera (Hamamatsu) and SimplePCI software (Compix) to collect Nomarski/DIC Z-stack data. We use ImageJ (free download; http://www.rsbweb.nih.gov/ij/), Vector Rose 3.0 (for Mac only; http://www.pazsoftware.com/VectorRose.html), and Microsoft Excel software for image processing and data analysis.

3. Methods

3.1. Live Cell Imaging and Quantification of PCP at the End of Gastrulation

1. Prior to tailbud stage (e.g., 80% epiboly) manually dechorionate 20–30 embryos of the desired phenotype using a pair of fine forceps. This is done carefully and in 30% Danieau's buffer to avoid damaging the embryos. Damaged embryos should be discarded and not used for imaging. Only use glass petri dishes and glass Pasteur pipets to prevent embryos from sticking (they can stick to plastic petri dishes and transfer pipets). Chorions and debris are removed from the dish and embryos are washed with fresh 30% Danieau's buffer. After dechorionation embryos in glass dishes are returned to the 28.5°C or 32°C dry incubator until they are at the proper developmental stage for imaging (see Note 1).

2. At least 1 h prior to embryos reaching the tailbud stage, place a 50 ml conical tube containing 30% Danieau's buffer in the 28.5°C dry incubator. At this time also place 1 ml aliquots of 0.8% LMP agarose in the 42°C heat block. We initially prepare and microwave 0.8% LMP agarose in a glass media bottle and then make aliquots in 1.5 ml microcentrifuge tubes and store these at room temperature.

3. At least 15 min prior to the tailbud stage use a microwave oven to melt the 3% agarose (we store and microwave this in a 500 ml Kimax-35 media bottle). Cool for 5–10 min then use a transfer pipet to add approximately 2.5–3 ml or 0.5 cm to several coverslip dishes (see Note 2 and Fig. 2a, b). Allow the agarose to completely harden at room temperature and then use a clean transfer pipet to poke 2–4 connected holes in the agarose that are directly centered over the glass coverslip (Fig. 2c). By squeezing the bulb prior to inserting the tip of the transfer pipet into the agarose you can aspirate the agarose chunks and thereby cleanly remove them from the coverslip dishes. The goal here is to generate an area over the coverslip that is devoid of 3% agarose. Also at this time make sure the

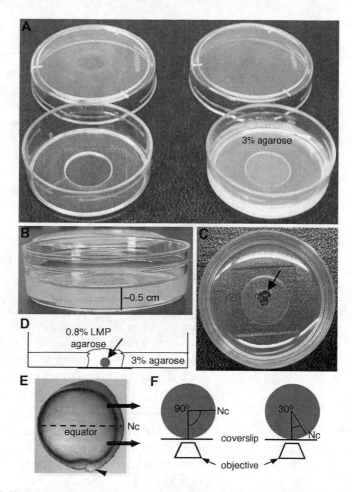

Fig. 2. Mounting tailbud stage embryos in agarose. (**a**) Coverslip dishes used for mounting procedure shown with and without 3% agarose. (**b**) Coverslip dish containing approximately 0.5 cm of 3% agarose. (**c**) Three holes generated in the 3% agarose using a transfer pipet (*arrow*). (**d**) Schematic drawing of embryo (*arrow*) mounted in 0.8% LMP agarose surrounded by 3% agarose. (**e**) Lateral view of a wild-type tailbud stage embryo with the equator centered in the middle. Here, the dorsal midline (notochord) is on the right (Nc) and the arrows denote the direction that you want to role the embryo to orient the notochord 30° from the objective. The *arrowhead* denotes the tailbud. (**f**) Schematic of embryo before (*left*) and after (*right*) orienting the dorsal midline (notochord, Nc) such that it is 30° from the objective.

microscope and computer are on and the heated stage is pre-warmed to 28.5°C.

4. When the dechorionated embryos reach tailbud stage you are ready to begin imaging (see Note 3). First, use a dissecting microscope to pick out 1–4 embryos that look good (i.e., have the phenotype you desire). Next, use a transfer pipet to fill the hole you previously generated in the 3% agarose with 42°C 0.8% LMP agarose (it is okay if you overflow the hole). Be careful that the LMP agarose is not much warmer than 42°C

otherwise embryo damage will occur. Quickly, using a glass Pasteur pipet, transfer 1–4 embryos to the 0.8% LMP agarose being careful to transfer as little 30% Danieau's buffer as possible. The embryos will settle to the bottom of the dish on top of the glass coverslip (see Fig. 2d). Using a dissecting microscope and fine forceps quickly orient the embryos such that you are looking at a lateral view with the equator at the middle (see Fig. 2e). Then role the embryos so their dorsal axes (i.e., the notochord or midline) are oriented towards the coverslip at a 30° angle (see Fig. 2f). While rolling the embryo, try to maintain the position of the equator near the center. Remember that the objective on an inverted microscope is below the glass coverslip thus you are mounting the embryos accordingly (see Note 4). You will have approximately 1 min before the 0.8% agarose hardens sufficiently such that further embryo manipulation is not possible. Therefore, it is best to attempt mounting only a single embryo until you become proficient (see Note 5). After the 0.8% agarose has hardened around the embryo(s) you can invert the coverslip dish and examine the embryo position using a dissecting microscope. When satisfied, add enough prewarmed 30% Danieau's buffer to cover the agarose and place the dish (without a lid) on the heated microscope stage.

5. First use either a 10× or 20× objective to center the embryo and find the approximate focal plane. Here you should also assess and document embryo orientation and quality before proceeding (i.e., take images and save as jpeg or tiff files). Make sure to acquire at least one image where the notochord is in focus (Fig. 3a). Then using a 40× objective and DIC optics bring the embryo into the desired focal plane for imaging (see Note 6). Endoderm consists of a scattered layer of large cells that form numerous fine membrane protrusions (Fig. 3b) and represents the deepest tissue layer visible above the amorphous yolk (15). Next is the multilayered mesoderm with the layer closest to the ectoderm appearing as tightly packed and smaller cells. The deepest layer of mesoderm is adjacent to the endoderm and the yolk. Deep mesodermal cells are morphologically distinct from the more superficial layers of mesoderm; they are larger and polarized along their dorsal path of directed migration (Fig. 3c). Above the mesoderm is the ectoderm with its characteristic cobblestone appearance (Fig. 3d). The last tissue to come into focus as you move away from the yolk is the enveloping layer consisting of large nucleated cells. Once satisfied with image quality you are ready to use the computer software to set the Z-stack parameters. We typically generate Z-stacks through the entire embryo (i.e., from the enveloping layer to the yolk) so that effects on each tissue layer can be analyzed if desired. Depending on microscope and objective capabilities a step size

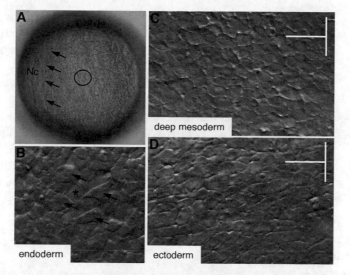

Fig. 3. Appearance of tissue layers using DIC microscopy. (**a**) ×20 magnification image of properly mounted and oriented embryo with the notochord (Nc, *arrows*) positioned vertically. The *circle* indicates the approximate area that will be imaged at ×40 magnification. (**b**) Endodermal cells (*arrows*) located adjacent to yolk (*asterisk*) imaged at ×40 magnification. (**c**) Polarized deep mesodermal cells imaged at ×40 magnification. (**d**) Polarized ectodermal cells imaged at ×40 magnification. In panels (**c**) and (**d**), the *horizontal white line* indicates the direction of cell polarization while the *vertical white line* denotes the orientation of the notochord (not shown).

of 0.5–1 µm works well (see Note 7). Save *Z*-stacks as movie files to facilitate importing them into ImageJ.

6. In ImageJ you must first open the file containing the whole embryo image (acquired using either a 10× or 20× objective). Rotate this image such that dorsal (i.e., notochord) is oriented vertically making note of exactly how many degrees of rotation were required (and clockwise or counter clockwise rotation direction). Next import the *Z*-stack movie file and rotate the entire stack of images just as the whole embryo image was rotated. The purpose of having a *Z*-stack is to ensure that the LWR's are calculated at the position of largest cellular diameter, something that will not be the same for every cell at a single focal plane. To keep track of the measured cells as you move up and down through the *Z*-stack (not possible with ImageJ), we simply tape a piece of plastic transparency to the computer monitor and mark each cell's position with a pen. Different colored pens can be used to denote the focal planes that were chosen to analyze different cells. To analyze cellular LWR and orientation you must select the "Set Measurements" function under Analyze and select "Fit Ellipse." Use the kidney bean selection icon on the toolbar to trace individual cell perimeters and for each cell select the "Measure" function under Analyze.

A new window will automatically open and will list in order values for each cell's major (length), minor (width), and angle (degrees from an imaginary horizontal line running across the screen). This data can be copy/pasted into Microsoft Excel for further manipulation including calculation of LWR (by dividing each cell's major value by the minor value) and statistical analyses. We typically display LWR data in the form of bar graphs with the Y-axis beginning with a LWR value of one (representing a perfect circle). Cellular orientation is analyzed using orientation analysis software such as Vector Rose 3.0 and can be presented in the form of circular histograms (see Note 8 and (5, 9, 11, 12) for examples).

4. Notes

1. In addition to 28.5°C you may want to grow some embryos at 32°C so that you can acquire images of multiple embryos. In this way you are staggering the time points when embryos reach tailbud stage (embryos grown several hours at 32°C will be at least 1 h further along in development than those grown at 28.5°C).

2. The exact volume is not important, as the goal here is to make the 3% agarose thicker than the diameter of a tailbud stage embryo. Prepare several dishes to account for performing multiple time-lapses and errors in embryo orientation.

3. Proper staging of embryos is critical especially when comparing control groups with mRNA/morpholino-injected embryos, as these tend to develop more slowly than uninjected controls. Therefore, the use of morphological criteria is essential and much more accurate than relying solely on hours post-fertilization. The tailbud stage is marked by the development of a posterior tailbud shortly after the end of epiboly (Fig. 2e) and the transient yolk plug closure stage (14).

4. The reasoning behind the recommended parameters for embryo orientation is that cells further than 20–30° from the dorsal embryonic axis (notochord) will be less polarized and thus more difficult to quantify while mesodermal cells closer to the dorsal midline are becoming presomitic mesoderm and will lose their polarity (note that ectodermal cells maintain polarity even near the dorsal midline, (5)).

5. Though challenging, we have found that mounting embryos in agarose is the best method to prevent rolling during imaging. This is extremely important when imaging at 40× magnification because even the slightest embryo movement will compromise data quality. The reason for mounting multiple

embryos is to expedite the experiment because not every embryo will be positioned correctly when viewed at high magnification.

6. It will take practice to become proficient at recognizing the different tissue layers of the embryo. Here I describe and show images of the tailbud stage embryo at approximately 30° from the dorsal midline or notochord. Previous studies have shown that ectodermal and deep mesodermal cells within the dorsal-lateral gastrula domain become polarized by the end of gastrulation (5, 8–12). Be aware that at different positions within the embryo and at different developmental stages the tissue layers will appear slightly different.

7. Though the focus of this chapter is the analysis of PCP, these same embryo mounting and orientation procedures are used to generate time-lapse movies of cell movement. After determining the tissue layer of interest, we set the microscope and imaging software to collect images at an interval of 30 s–1 min for typically 30 min. Here, we often simultaneously collect Z-stack data through the entire focal plane of the tissue layer. In this way we can follow the behavior of a cell even if it moves slightly above or below a given focal plane.

8. The angle values created in ImageJ range from 0 to 180 with values nearer 0 and 180 representing cells whose long axes (the major value) are parallel with a horizontal line running across the screen (or perpendicular to a vertical line). Since the embryo was rotated such that the notochord was oriented vertically, these angle values accurately quantify cellular orientation in relation to the path of directed migration (i.e., towards the notochord). Typically, a cell is classified as being mediolaterally aligned (i.e., perpendicular to the notochord) if its long axis is oriented ±20° in relation to the notochord (5, 9, 12). Thus, based on ImageJ's calculations, this will include cells that produce angle values of 0–20 and 160–180. By using a program such as Vector Rose 3.0 the angle of each cell from the entire dataset (i.e., hundreds of cells from multiple embryos) can be displayed in a single circular histogram.

Acknowledgments

I thank my colleagues Diane Sepich and Lila Solnica-Krezel for their investment in my training in all methods concerning zebrafish gastrulation. Work in the Jessen lab is supported by grants from the American Cancer Society (RSG-09-281-01 DDC) and National Science Foundation (IOS 0950849).

References

1. Wong, L. L., and Adler, P. N. (1993) Tissue polarity genes of Drosophila regulate the subcellular location for prehair initiation in pupal wing cells, *J Cell Biol 123*, 209–221.

2. Adler, P. N. (2002) Planar signaling and morphogenesis in Drosophila, *Dev Cell 2*, 525–535.

3. Hammerschmidt, M., Pelegri, F., Mullins, M. C., Kane, D. A., Brand, M., van Eeden, F. J., Furutani-Seiki, M., Granato, M., Haffter, P., Heisenberg, C. P., Jiang, Y. J., Kelsh, R. N., Odenthal, J., Warga, R. M., and Nusslein-Volhard, C. (1996) Mutations affecting morphogenesis during gastrulation and tail formation in the zebrafish, Danio rerio, *Development 123*, 143–151.

4. Solnica-Krezel, L., Stemple, D. L., Mountcastle-Shah, E., Rangini, Z., Neuhauss, S. C., Malicki, J., Schier, A. F., Stainier, D. Y., Zwartkruis, F., Abdelilah, S., and Driever, W. (1996) Mutations affecting cell fates and cellular rearrangements during gastrulation in zebrafish, *Development 123*, 67–80.

5. Jessen, J. R., Topczewski, J., Bingham, S., Sepich, D. S., Marlow, F., Chandrasekhar, A., and Solnica-Krezel, L. (2002) Zebrafish trilobite identifies new roles for Strabismus in gastrulation and neuronal movements, *Nat Cell Biol 4*, 610–615.

6. Solnica-Krezel, L., and Cooper, M. S. (2002) Cellular and genetic mechanisms of convergence and extension, *Results Probl Cell Differ 40*, 136–165.

7. Mlodzik, M. (2002) Planar cell polarization: do the same mechanisms regulate Drosophila tissue polarity and vertebrate gastrulation?, *Trends Genet 18*, 564–571.

8. Concha, M. L., and Adams, R. J. (1998) Oriented cell divisions and cellular morphogenesis in the zebrafish gastrula and neurula: a time-lapse analysis, *Development 125*, 983–994.

9. Coyle, R. C., Latimer, A., and Jessen, J. R. (2008) Membrane-type 1 matrix metalloproteinase regulates cell migration during zebrafish gastrulation: evidence for an interaction with non-canonical Wnt signaling, *Exp Cell Res 314*, 2150–2162.

10. Lin, F., Sepich, D. S., Chen, S., Topczewski, J., Yin, C., Solnica-Krezel, L., and Hamm, H. (2005) Essential roles of G{alpha}12/13 signaling in distinct cell behaviors driving zebrafish convergence and extension gastrulation movements, *J Cell Biol 169*, 777–787.

11. Myers, D. C., Sepich, D. S., and Solnica-Krezel, L. (2002) Bmp activity gradient regulates convergent extension during zebrafish gastrulation, *Dev Biol 243*, 81–98.

12. Topczewski, J., Sepich, D. S., Myers, D. C., Walker, C., Amores, A., Lele, Z., Hammerschmidt, M., Postlethwait, J., and Solnica-Krezel, L. (2001) The zebrafish glypican knypek controls cell polarity during gastrulation movements of convergent extension, *Dev Cell 1*, 251–264.

13. Sepich, D. S., and Solnica-Krezel, L. (2005) Analysis of cell movements in zebrafish embryos: morphometrics and measuring movement of labeled cell populations in vivo, *Methods Mol Biol 294*, 211–233.

14. Kimmel, C. B., Ballard, W. W., Kimmel, S. R., Ullmann, B., and Schilling, T. F. (1995) Stages of embryonic development of the zebrafish, *Dev Dyn 203*, 253–310.

15. Warga, R. M., and Nusslein-Volhard, C. (1999) Origin and development of the zebrafish endoderm, *Development 126*, 827–838.

Chapter 7

Wnt/Planar Cell Polarity Signaling in the Regulation of Convergent Extension Movements During *Xenopus* Gastrulation

Gun-Hwa Kim, Edmond Changkyun Park, and Jin-Kwan Han

Abstract

The Wnt/planar cell polarity (PCP) signaling pathway plays a critical role in wing, eye, neural tube defects, and sensory bristle development of *Drosophila* and vertebrate development. Recently, the Wnt/PCP pathway has been known to regulate convergent extension (CE) movements that are essential for establishing the three germ layers and body axis during early vertebrate development. Here, we describe detailed practical procedures required for the particular studies in *Xenopus* CE movements

Key words: Convergent extension (CE) movements, Planar cell polarity (PCP), Wnt pathway, *Xenopus* development

1. Introduction

Morphogenetic movements in gastrulation are critical for establishing basic germ layers and the body axis during early vertebrate development. The major driving forces for this process include convergent extension (CE) movements, by which cells polarize and elongate along the mediolateral axis and intercalate toward the midline (convergence), leading to extension of the tissue along the anterior/posterior axis (1, 2). Although the precise molecular mechanisms of CE movements are not clearly understood, the Wnt/planar cell polarity (PCP) pathway is known to be important in the control of CE movements (3, 4).

The Wnt/PCP pathway is initiated by binding of the Wnt ligand Wnt11 to the Wnt receptor Frizzled 7 (Fz7) on the surface of target cells. In turn, activated Fz, a seven transmembrane domain

Kursad Turksen (ed.), *Planar Cell Polarity: Methods and Protocols*, Methods in Molecular Biology, vol. 839,
DOI 10.1007/978-1-61779-510-7_7, © Springer Science+Business Media, LLC 2012

protein receptor, recruits the cytosolic protein Disheveled (Dsh) (5, 6). The Fz7/Dsh complex activates the β-arrestin 2 (βarr2), Daam1, RhoA, Rac1, Rho-associated kinase α (ROKα), and Jun N-terminal kinase (JNK) (2, 3, 7, 8).

In addition to CE movements, Fz7 has been reported to regulate tissue separation. During *Xenopus* gastrulation, the involuted mesendoderm is separated from the ectoderm by a process called tissue separation. As a result, Brachet's cleft is generated between the mesendoderm and ectoderm (9). The separation behavior between the mesendoderm and ectoderm has been known to be regulated by Fz7 receptor function (10). The regulation of tissue separation can easily be studied by analyzing the Brachet's cleft formation in the gastrula embryos.

2. Materials

2.1. In Vitro Fertilization and Dejelly

- *Xenopus laevis* (Nasco, http://www.enasco.com/xenopus/).

- Human Chorionic Gonadotropin (hCG, Sigma): Prepare in a concentration of 1,000 IU/ml in sterile water. It should be stored at 4°C and used within 1 month.

- 10× Marc's Modified Ringers (10× MMR): 1 M NaCl, 20 mM KCl, 10 mM $MgSO_4$, 20 mM $CaCl_2$, 50 mM HEPES, 1 mM EDTA. Before use, dilute 1:10 (1× MMR) and adjust to pH 7.4.

- 1× High-salt Modified Barth's Saline (1× High-salt MBS): 108 mM NaCl, 1 mM KCl, 0.7 mM $CaCl_2$, 1 mM $MgSO_4$, 5 mM HEPES, 2.5 mM $NaHCO_3$. Adjust to pH 7.4.

- Calf serum (Invitrogen).

- Gentamycin.

- 1× Modified Frog Ringers (1× MR): 0.1 M NaCl, 1.8 mM KCl, 2 mM $CaCl_2$, 1 mM $MgCl_2$, 5 mM HEPES. Before use, dilute 1:10 (0.1× MR) or 1:3 (0.33× MR) and adjust to pH 7.4.

- Dejellying solution: 2% cystine hydrochloride in 0.33× MR, pH 7.8–8.0.

2.2. Microinjections

- Injector (PIL-100, Harvard Apparatus) and manipulator.

- Glass capillary needles: Pull glass needles with a micropipette puller (Narishige, PC-10) and grind the end tip of the needle with a micropipette grinder (Narishige, EG-44).

- Petri dishes (with or without netting): Prepare small discs of polyprophylene netting (Spectra/Mesh, 146410) and glue the net onto the 60mm dish with paraffin.

– Injection solutions of RNA or MO.

– 4% Ficoll in 0.33× MR.

2.3. Dorsal Marginal Zone Elongation Assay

– 1× Steinberg's solution: 60 mM NaCl, 0.67 mM KCl, 0.34 mM Ca(NO$_3$)$_2$, 0.83 mM MgSO$_4$, 10 mM HEPES. Adjust to pH 7.4.

– 2% Agarose-coated dish: Coat the bottom of 60 mm petri dish with 2% agarose in 1× Steinberg's solution.

– Dorsal marginal zone (DMZ) culture media: 50 μg/ml gentamycin, 2.5 mg/ml spectromycin sulfate, and 1 mg/ml BSA in 1× Steinberg's solution.

2.4. Glutathione S-Transferase Pull-Down for RhoA Activity Assay

– GST-RBD (RhoA binding domain) protein (For purification of GST-RBD protein, please refer (11, 12)).

– Glutathione S-Transferase (GST)-beads (Immobilized glutathione agarose resin, Pierce).

– RIPA buffer: 150 mM NaCl, 50 mM Tris–HCl (pH 8.0), 1% Triton X-100, 1% sodium deoxycholate, 2 mM EDTA, 25 mM beta-glycerophosphatate, 100 mM NaF, 1 mM Na$_3$VO$_4$, 2 mM DTT, and protease inhibitors.

– GST-fish buffer: 10% glycerol, 50 mM Tris–HCl (pH 7.4), 100 mM NaCl, 2 mM MgCl$_2$, 1% Igepal CA 630.

2.5. SDS-Polyacrylamide Gel Electrophoresis

– BioRad Mini-Protean System.

– 1.5 M Tris–HCl, pH 8.8.

– 1 M Tris–HCl, pH 6.8.

– 30% Acrylamide/bis solution (29: 1). This is a neurotoxin when unpolymerized and so care should be taken not to receive exposure.

– N,N,N,N'-Tetramethyl-ethylenediamine (TEMED).

– Ammonium persulfate: prepare 10% solution in water and immediately freeze in single use (200 μl) aliquots at –20°C.

– 12% resolving gel solution: (for 10 ml) 3.3 ml H$_2$O, 4.0 ml of 30% acrylamide mix, 2.5 ml of 1.5 M Tris–HCl (pH 8.8), 0.1 ml of 10% SDS, 0.1 ml of 10% APS, 0.004 ml of TEMED.

– 5% stacking gel solution: (for 5 ml) 2.7 ml H$_2$O, 0.67 ml of 30% acrylamide mix, 0.5 ml of 1.0 M Tris–HCl (pH 6.8), 0.04 ml of 10% SDS, 0.04 ml of 10% APS, 0.004 ml of TEMED.

– 100% isopropanol. Use the top layer.

– Running buffer (5×): 25 mM Tris, 250 mM glycine, 0.1% (w/v) SDS.

– Prestained molecular weight markers (Fermentas, #SM1811).

2.6. Western Blotting

- Nitrocellulose membrane (Wattman, 0.2 μm pore size).
- 3 MM paper.
- Transfer buffer (1× for semi-dry): 25 mM Tris, 250 mM Glycine, 15% methanol.
- Semi-dry transfer unit (Hoefer).
- 10× TBST (Tris-buffered Saline Tween-20): 1 M Tris–HCl, 1.5 M NaCl, 1% Tween 20, pH 7.4). Dilute 1:10 (1× TBST) before use.
- Blocking buffer: 5% (w/v) nonfat dry milk in TBST.
- Primary antibody dilution buffer: TBST supplemented with 3% (w/v) fraction V bovine serum albumen (BSA).
- Primary antibody: phospho-JNK (Cell Signaling, #9251), JNK (Cell Signaling, #9252), RhoA (Cell Signaling, #2117), actin (Santa Cruz, sc-1616).
- Secondary antibody: Anti-rabbit (Sigma, A0545) and -mouse (Sigma, F5387) IgG conjugated to horse radish peroxidase (HRP).
- Enhanced chemiluminescent (ECL) prime Western blotting reagents (GE healthcare, RPN2232SK).
- LAS-4000 (Fujifilm).

2.7. Scanning Electron Microscopy for Observing Brachet's Cleft

- 0.1 M Sodium cacodylate buffer, pH 7.4.
- 2.5% Glutaraldehyde in 0.1 M sodium cacodylate buffer, pH 7.4.
- 1% Osmium tetroxide, pH 7.4.
- 25, 50, 70, 85, and 100% Ethanol.
- Double-sided carbon tape.
- 2% Agarose-coated dish (60 mm).

3. Method

3.1. Inducing Ovulation, Collecting Eggs, and In Vitro Fertilization

1. For inducing ovulation, inject 600–800 IU of hCG into the dorsal lymph sac of female *Xenopus laevis* and keep the frog in dark condition at 20°C overnight. Generally, it takes about 12–15 h for frogs to start ovulating (refer (13) for detailed *Xenopus* handling method).

2. When female frog starts to ovulate, place the frog into the container with 1× MMR solution instead of tap water to collect eggs (see Note 1).

3. During collecting eggs, isolate the testes by sacrificing a male frog. Store isolated testes in 20% serum in 1× high-salt MBS with 0.05 mg/ml gentamycin at 4°C (refer (13) for detailed method of isolating testes).

4. Once the frog ovulated as much as you need, transfer the eggs to 60 mm dish using transfer pipette, rinse carefully with 0.33× MR, and remove all the buffer from the eggs.

5. Crush a small piece of the testes in 0.33× MR and mix the sperm with the eggs (see Note 2). Keep the eggs at 20°C. Usually, the eggs are fertilized within 30 min.

6. Prepare fresh dejellying solution during in vitro fertilization.

7. Once the eggs are fertilized, change the sperm suspension with dejellying solution and gently stir the eggs until the jelly membranes are completely removed from the eggs. When jelly is removed, eggs are getting to pack together closely. It is usually complete in 5 min.

8. After dejellying, rinse the fertilized eggs at least ten times with 0.33× MR to remove the cysteine and transfer the embryos in a clean dish with 0.33× MR.

3.2. Microinjection and Phenotype Analysis

1. Under the dissecting microscope, put injection solution onto a piece of parafilm and load the target RNA or MO into glass capillary needle.

2. 4-Cell stage embryos are transferred into the injection dish (with netting) filled with 4% Ficoll solution.

3. Embryos are injected with mRNA or MO into both sites of dorsal blastomeres at equatorial region (see Note 3).

4. Transfer injected embryos into new dishes with 4% Ficoll in 0.33× MR and culture at 14°C or 20°C.

5. Change the culture media to 0.1× MR at stage 8 and culture until the purposed stage. Examples of the CE defective phenotypes are shown in Fig. 1 (see Note 4).

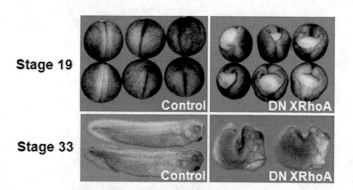

Fig. 1. The phenotypic defects in convergent extension movements. Dorsal two blastomeres of four-cell stage embryos were injected with dominant negative (DN) *Xenopus* RhoA (XRhoA) mRNA (200 pg). DN XRhoA-injected embryos exhibited a failure of neural tube closure and exposure of an endodermal mass on the dorsal side at late neurula stages (stage 19) and kinked and shortened *A–P* axis at tailbud stages (stage 33). This figure is referenced in and modified from ref. 8.

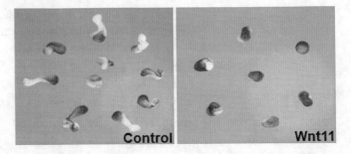

Fig. 2. Dorsal marginal zone (DMZ) elongation assay. Two blastomeres of four-cell stage embryos were injected at the dorsal equatorial region with Wnt11 (500 pg). Dorsal marginal zone (DMZ) explants from control embryos elongated, but the explants expressing Wnt11 were significantly inhibited.

3.3. DMZ Elongation Assay

1. Place the stage 10.5 embryos in 1× Steinberg's solution on 2% agarose-coated dish.

2. Prepare two sharp forceps for DMZ tissue dissection to remove vitelline envelope of the embryos. Pick up the envelope on ventral side by one forceps and tear the envelope from embryo using another forceps.

3. Put the ventral region of embryos upward and animal pole region downward. Press the embryos lightly as oval-globe shape to control the overturned embryos.

4. Cut the dorsal marginal tissue of embryos. First, by each tip of a forceps, pierce the embryo nearby both ends of dorsal lip. Cut the embryos along two parallel lines in the vertical direction of dorsal lip at the same time. After turning the overturned embryo upside down again, make a parallel section with dorsal lip at the 1/4 position of diameter. Peel off the dorsal marginal tissues to the dorsal lip. If they don't get separated easily, split them with compulsion by forceps. Cut the dorsal tissues below dorsal lip. Lay the dorsal tissue and the embryo apart carefully. Using the tips of forceps, remove the head mesoderm cell on explants completely without wound.

5. To open face explants, wait until the explants become globe shaped and transfer the explants to new 2% agarose-coated dish with DMZ culture media. An example of the results produced is shown in Fig. 2.

3.4. RhoA and JNK Activity Assay

[GST Pull Down]

1. DMZ tissues are isolated at stage 10.5 and cultured in culture media.

2. At stage 12, DMZ tissues are lysed in RIPA buffer (50 DMZ tissues in 500 μl RIPA buffer) and centrifuged at 13,000×g at 4°C for 15 min.

3. Cell lysates are incubated with GST-RBD beads at 4°C for 3 h (see Note 5). Leave cell lysates for detecting total RhoA protein and JNK activity.

4. Lysate and GST-bead mixture are washed four times with 2 volumes of GST-fish buffer and are eluted by cooking at 95°C within protein loading buffer.

[SDS-PAGE]

1. Assemble BioRad Mini-protean system and prepare 10 ml of 12% resolving gel solution, pour the resolving gel mixture into the plates, and overlay with isopropanol (Leave the space for stacking gel). The gel should polymerize within about 30 min.

2. Remove the isopropanol and rinse the top of the gel twice with water.

3. Prepare 5 ml of 5% stacking gel solution, pour the stacking gel mixture, and insert the comb. The stacking gel should polymerize within 20 min.

4. Once the stacking gel polymerizes, place the gel plates into the running camber and fill up the gel running buffer into the camber.

5. Remove the comb and wash the lane by pipetting.

6. Load the samples and run the gel at 10 mA through the stacking gel and 20 mA through the resolving gel.

[Western Blotting]

1. Once the samples are separated by SDS-polyacrylamide gel electrophoresis (SDS-PAGE), remove the gel from the plates and transfer into a tray containing transfer buffer.

2. 3 MM papers and nitrocellulose membrane are prepared in the same size with the gel and wetted with transfer buffer.

3. Assemble the membrane transfer sheets (from the bottom, two sheets of wetted 3 MM paper, NC membrane, gel, two sheets wetted of 3 MM paper) and transfer the protein from the gel to membrane by using Hoefer semi-dry transfer unit.

4. Once the transfer is complete, incubate the membrane in the blocking solution for 1 h at room temperature on a rocking platform.

5. Discard the blocking solution and incubate with primary antibody (anti-RhoA, start from 1:1,000) in blocking solution for 2 h at room temperature on a rocking platform.

6. Remove the primary antibody solution and wash the membrane three times with TBST for 3 min.

7. Incubate with secondary antibody (start from 1:10,000) in blocking solution for 1 hour at room temperature on a rocking platform.

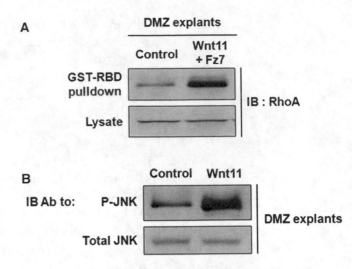

Fig. 3. Analysis of RhoA and JNK activity assay in DMZ tissues. Two blastomeres of four-cell stage embryos were injected at the dorsal marginal region with indicated mRNAs. The DMZ explants were dissected at stage 10.5 and cultured until stage 12. (**a**) Wnt11 and Fz7 activated RhoA in DMZ tissues during gastrulation. GTP-bound RhoA in DMZ lysates was precipitated using GST-RBD and visualized by immunoblotting with anti-RhoA antibody. (**b**) Wnt11 increased JNK phosphorylation in CE movements. The explant lysates were blotted with anti-phospho JNK and anti-JNK antibodies.

8. Remove the secondary antibody solution and wash the membrane five times with TBST for 5 min.

9. During the final wash, warm up the ECL reagent to room temperature.

10. Spray the ECL reagent onto the membrane and detect the signal by using the LAS-4000 (see Note 6).

11. For detecting total RhoA, perform the SDS-PAGE and Western blotting again with whole cell lysates.

12. For JNK activity assay, cell lysates are examined by Western blotting with JNK and phospho-JNK antibodies. JNK antibody shows total amount of JNK protein and phospho-JNK antibody represents active JNK proteins (phosphorylated). An example of JNK and RhoA activity assay produced is shown in Fig. 3.

3.5. SEM Sample Preparation for Observing Brachet's Cleft

1. Fix the *Xenopus* embryo at stage 11.5 with 2.5% glutaraldehyde in 0.1 M sodium cacodylate buffer for 4 h at room temperature or at 4°C overnight.

2. Rinse the embryo three times with 0.1 M sodium cacodylate buffer.

3. Place the fixed embryo sample on the agarose gel bed and bisect sagittally through the dorsal midline with a vibratome blade.

4. Immerse the sample in 1% osmium tetroxide for 1 h at room temperature in dark condition.

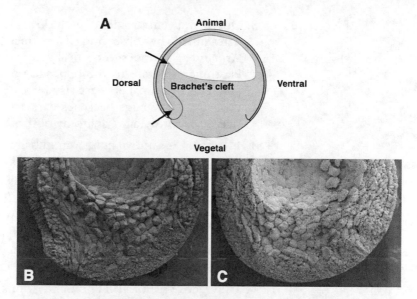

Fig. 4. Analysis of Brachet's cleft by SEM. (**a**) Illustration of Brachet's cleft at stage 11. The length of Brachet's cleft is indicated with *arrows*. (**b**) SEM image from the control embryo. Control embryo shows a normal Brachet's cleft formation in dorsal side. (**b**) SEM image from Fz7 deficient embryo. Brachet's cleft in dorsal side is not visible since tissue separation between the mesendoderm and ectoderm is inhibited by loss-of-Fz7 function. In contrast, cleft in ventral side is normally formed as does in the control embryo.

5. Rinse the sample three times with 0.1 M sodium cacodylate buffer.

6. Dehydrate the sample serially with ethanol (10 min in 25% ethanol, 10 min in 50% ethanol, 10 min in 70% ethanol, 10 min in 85% ethanol, and then 10 min in 100% ethanol) and completely air-dry the sample.

7. Mount the dried sample onto metal stub with double-sided carbon tape.

8. Coat over the sample with gold in a thin layer using an automated sputter coater just before observing scanning electron microscopy (SEM) image. Examples of the SEM images are shown in Fig. 4.

4. Notes

1. Container with mesh on the bottom, whose pore is little bit bigger than the size of eggs, makes it much easier to collect egg and keep the egg safer.

2. Optionally, before crushing the testes, rub a small piece of the testes all over the eggs. This may help to increase fertilization rate when the testes are isolated days ago.

3. Dorsal and ventral sides of early *Xenopus* embryos can be easily determined by observing its pigment. Upon fertilization, *Xenopus* eggs undergo cortical rotation, by which the vegetal blastomeres rotate to the opposite point of sperm entry and pigments in animal pole are also relocated. As a result, the ventral half of the eggs becomes darker than dorsal half (Please refer (14) for detailed anatomy of the *Xenopus* embryos).

4. Effects on CE movements can initially be determined by analyzing injection phenotypes. When CE movements are inhibited, embryos exhibited opened blastopore at late neurula stages and kinked and shortened axis at tailbud stages.

5. To prepare GST-RBD bead for pulldown, add GST-RBD protein (20 μg) to GST-beads (1/10 volume of cell lysates) and incubate at 4°C for 15 min. Wash the GST-beads two times with 2 volumes of RIPA buffer before use.

6. Only active RhoA (GTP-bound) can be detected from this method. The amount of RBD-bound RhoA should be normalized to the total RhoA in cell lysates for the comparison of RhoA activity.

Acknowledgments

We acknowledge Mi-Jung Kim and other members of our laboratory for reading this manuscript and providing constructive criticism. This work was supported by the National Research Foundation of Korea (KRF) grant (No. 20090092829) and K-MeP project (T31130) of Korea Basic Science Institute (KBSI).

References

1. Veeman MT, Axelrod JD, Moon RT. (2003) A second canon. Functions and mechanisms of beta-catenin-independent Wnt signaling. *Dev Cell* Sep;5(3):367–77.

2. Wallingford JB, Fraser SE, Harland RM. (2002) Convergent extension: the molecular control of polarized cell movement during embryonic development. *Dev Cell* Jun;2(6):695–706.

3. Myers DC, Sepich DS, Solnica-Krezel L. (2002) Convergence and extension in vertebrate gastrulae: cell movements according to or in search of identity? *Trends Genet* Sep;18(9):447–55.

4. Wallingford JB, Habas R. (2005) The developmental biology of Dishevelled: an enigmatic protein governing cell fate and cell polarity. *Development* Oct;132(20):4421–36.

5. Tada M, Smith JC. (2000) Xwnt11 is a target of Xenopus Brachyury: regulation of gastrulation movements via Dishevelled, but not through the canonical Wnt pathway. *Development* May;127(10):2227–38.

6. Djiane A, Riou J, Umbhauer M, Boucaut J, Shi D. (2000) Role of frizzled 7 in the regulation of convergent extension movements during gastrulation in Xenopus laevis. *Development* Jul;127(14):3091–100.

7. Kim GH, Han JK. (2007) Essential role for beta-arrestin 2 in the regulation of Xenopus convergent extension movements. *EMBO J* May 16;26(10):2513–26.

8. Kim GH, Han JK. (2005) JNK and ROKalpha function in the noncanonical Wnt/RhoA signaling pathway to regulate Xenopus convergent

extension movements. *Dev Dyn* Apr;232(4): 958–68.

9. Wacker S, Grimm K, Joos T, Winklbauer R. (2000) Development and control of tissue separation at gastrulation in Xenopus. *Dev Biol* Aug 15;224(2):428–39.

10. Winklbauer R, Medina A, Swain RK, Steinbeisser H. (2001) Frizzled-7 signalling controls tissue separation during Xenopus gastrulation. *Nature* Oct 25;413(6858): 856–60.

11. Ren XD, Kiosses WB, Schwartz MA. (1999) Regulation of the small GTP-binding protein Rho by cell adhesion and the cytoskeleton. *EMBO J* Feb 1;18(3):578–85.

12. Park E, Kim GH, Choi SC, Han JK. (2006) Role of PKA as a negative regulator of PCP signaling pathway during Xenopus gastrulation movements. *Dev Biol* Apr 15;292(2):344–57.

13. Sive HL, Grainger RM, Harland RM. (2000) Early Development of Xenopus laevis: A Laboratory Manual. Cold Spring Harbor Laboratory Press.

14. Nieuwkoop PD, Faber J. (1967) Normal table of Xenopus laevis (Daudin). Amsterdam: North Holland.

Chapter 8

Using 32-Cell Stage *Xenopus* Embryos to Probe PCP Signaling

Hyun-Shik Lee, Sergei Y. Sokol, Sally A. Moody, and Ira O. Daar

Abstract

Use of loss-of function (via antisense Morpholino oligonucleotides (MOs)) or over-expression of proteins in epithelial cells during early embryogenesis of *Xenopus* embryos, can be a powerful tool to understand how signaling molecules can affect developmental events. The techniques described here are useful for examining the roles of proteins in cell–cell adhesion, and planar cell polarity (PCP) signaling in cell movement. We describe how to target specific regions within the embryos by injecting an RNA encoding a tracer molecule along with RNA encoding your protein of interest or an antisense MO to knock-down a particular protein within a specific blastomere of the embryo. Effects on cell–cell adhesion, cell movement, and endogenous or exogenous protein localization can be assessed at later stages in specific targeted tissues using fluorescent microscopy and immunolocalization.

Key words: Planar cell polarity, *Xenopus*, Immunofluorescence, Blastomeres, Cell movement

1. Introduction

The polarized or directional orientation of cells and the migration of cells are controlled by the planar cell polarity (PCP) signaling pathway, which is critical for many developmental processes including the apical-basal polarity of epithelial tissue, convergent-extension movements during gastrulation and neurulation, and even the orientation of inner ear sensory cells and hair follicles (1).

In vertebrates, mutations in elements of the PCP pathway can cause various developmental abnormalities affecting morphogenesis in the neural tube, kidney, heart, and sensory organs (2). Disruptions in PCP components may lead to more invasive and metastatic properties of cells, and as such, may be considered to normally

Kursad Turksen (ed.), *Planar Cell Polarity: Methods and Protocols*, Methods in Molecular Biology, vol. 839,
DOI 10.1007/978-1-61779-510-7_8, © Springer Science+Business Media, LLC 2012

exhibit tumor suppressive functions (3). Thus, PCP signaling and the functional outcomes of this signaling pathway have been areas of intense focus.

The plasma membrane of epithelial cells is a wonderful example of cell polarity that is divided into two domains. This membrane consists of the apical domain facing the external environment and the basolateral domain in contact with the internal milieu. Epithelial cells have four different physical junctional structures, tight junctions, adherens junctions, gap junctions, and desmosomes. Of particular interest, the tight junctions separate the apical and basolateral domain in each cell, which is important to maintain cell polarity (4). The general features of cell polarity have been well defined in cultured epithelial cells (5, 6); however, polarity plays a critical role in maintaining an orderly process for proper early embryonic development and tissue morphogenesis (7, 8).

We have found the 32-cell stage *Xenopus* embryo to be a very useful system for examining how proteins that interact with the PCP pathway or represent essential components of the pathway can affect developmental processes. An advantage of this system is its well-characterized and consistent cell fate map, which allows a cell's lineage to be easily traced during experiments (9). Translation of specific endogenous proteins (affecting PCP signaling) can be inhibited, mutant proteins can be ectopically expressed in embryos with great facility and developmental effects can be examined within 2–3 days. Thus, signal transduction and differentiation processes can be assessed morphologically, histologically, as well as biochemically in a developing vertebrate. Fluorescence microscopy is an extremely useful methodology to examine the in vivo localization of endogenous proteins in whole organisms, and is particularly informative when examining the effects of knock-down or over-expression of specific gene products, such as cell adhesion molecules.

Using the epithelial cells of early stage *Xenopus* embryos, we have recently shown that loss-of function of proteins or over-expressing proteins that interact with Par polarity complex can disrupt cell–cell contacts and tight junctions (10). In addition to the roles of proteins in cell–cell adhesion, one can examine the role of proteins affecting PCP signaling in cell movement. Using the methods outlined here, we recently discovered how a specific protein (i.e., ephrinB1) can interact with the PCP pathway to regulate the movement of retinal progenitor cells into the eye field (11–13). Effective use of this system entails targeting specific regions within the embryos by injecting an mRNA encoding a tracer molecule along with mRNA encoding your protein of interest or an antisense MO against your protein into a specific blastomere. Effects on cell–cell adhesion, cell movement, and endogenous or exogenous protein localization can be assessed at later stages in specific targeted tissues using fluorescent microscopy and immunolocalization.

2. Materials

2.1. Equipment

1. Poly (methyl methacrylate) (PMMA) chamber (tank): Fig. 1.

2. Programmable injectors: PLI-100 Pico-Injector, Medical Systems Corp (Greenvale, NY), Narishige IM300 Microinjector (Greenvale, NY).

3. Programmable micropipette puller: horizontal puller (Greenvale, NY).

4. Stereomicroscopes.

5. Micromanipulators with mounted on a magnetic base secured to a steel plate.

6. Injection dishes: Nitex mesh (Fisher Scientific #8-670-176) on the 35 mm petri dishes.

7. Cryostat with microtome blade.

8. Nutators.

9. Superfrost slides.

10. Cover slips.

11. Fine sharpened dissecting needles.

12. Glass scintillation vials.

13. Embedding molds: Tissue-Tek Cryomold #62534-10.

14. PAP pens.

15. Slide holders.

16. Slide reservoirs.

17. Humidity chambers.

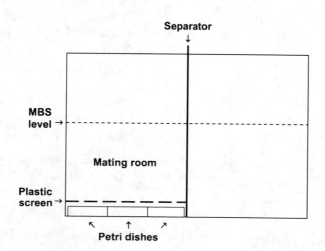

Fig. 1. Schematic diagram for PMMA chamber. Cartoon depicting side view of the chamber needed for natural mating.

2.2. Obtaining Embryos

1. HCG: human chorionic gonadotropin made with sterile water at a concentration of 2 IU/μl.

2. 0.1× MBS (Modified Barth's Solution): 8.8 mM NaCl; 0.1 mM KCl; 0.1 mM $MgSO_4$; 0.5 mM HEPES, pH 7.8; 0.25 mM $NaHCO_3$, 0.07 mM $CaCl_2$ in 1 l of distilled water.

3. 0.5× MBS: 44 mM NaCl; 0.5 mM KCl; 0.5 mM $MgSO_4$; 2.5 mM HEPES, pH 7.8; 1.25 mM $NaHCO_3$, 0.35 mM $CaCl_2$ in 1 l of distilled water.

4. 1× MBS: 88 mM NaCl; 1.0 mM KCl; 1.0 mM $MgSO_4$; 5.0 mM HEPES, pH 7.8; 2.5 mM $NaHCO_3$, 0.7 mM $CaCl_2$ in 1 l of distilled water.

5. Dejellying solutions: 2% L-cysteine in 0.5× MBS pH to 8.1 by adding 10 N NaOH.

6. Ficoll solution: 3% ficoll in 0.5× MBS.

2.3. Injections into a Specific Blastomere

1. mRNAs: GFP (Green Fluorescence Protein) or β-galactosidase as lineage tracers; gene of interest.

2. MOs: antisense morpholino oligonucleotides. These can be purchased with a fluorescent tag and thereby also act as a lineage tracer.

2.4. Fixation and Histochemical Reactions

2.4.1. Fixation and Histochemical Reactions for β-Galactosidase

1. 0.7× PBS (Phosphate-buffered Saline): 0.179 g of NaH_2PO_4·H_2O; 0.835 g of Na_2HPO_4·H_2O; 7.154 g of NaCl in 1 l of distilled water. Adjust pH to 7.4.

2. β-Galactosidase fixative: 2% Formaldehyde, 0.2% Glutaraldehyde, 0.02% Triton X-100, and 0.01% Sodium Deoxycholate in 0.7× PBS.

3. β-Galactosidase staining solution: 5 mM $K_3Fe(CN)_6$, 5 mM $K_4Fe(CN)_6$, and 1 mg/ml of X-gal (or Red-gal), 1 mM $MgCl_2$ in 0.7× PBS.

4. MEMFA: 0.1 M MOPS (pH 7.4); 2 mM EGTA, 1 mM $MgSO_4$; 3.7% formaldehyde.

5. Bleaching solution: 1% H_2O_2; 5% formamide; 0.5× SSC (standard saline citrate).

2.4.2. Fixation, Embedding, and Sections for GFP

1. 1× PBS: 0.256 g of NaH_2PO_4·H_2O; 1.194 g of Na_2HPO_4·H_2O; 10.22 g of NaCl in 1 l of distilled water. Adjust pH to 7.4.

2. GFP fixative: 4% Paraformaldehyde, 0.9 g of NaCl, 40 ml of 0.2 M Na_2HPO_4, 10 ml of 0.2 M NaH_2PO_4 in 100 ml of 1× PBS.

3. OCT compound: Tissue-Tek #4583.

4. Vectashield: Vectorlabs #H-1400.

5. Nail polish.

1. 0.1× MBS (Modified Barth's Solution): 8.8 mM NaCl; 0.1 mM KCl; 0.1 mM MgSO$_4$; 0.5 mM HEPES, pH 7.8; 0.25 mM NaHCO$_3$ in 1 l of distilled water.

2. Dent's fixative: 4 volumes of methanol; 1 volume of dimethyl sulfoxide (DMSO).

3. 1× PBS.

4. 15% Gelatin/sucrose solution: 16.67 ml of fish gelatin (45% stock); 7.5 g of sucrose in 50 ml of distilled water.

5. Sodium azide: 10% (w/v) stock.

6. OCT compound: Tissue-Tek #4583.

7. Acetone.

8. Image-iT signal enhancer: Invitrogen #I136933.

9. Blocking solution: 1% BSA; 5% heat-inactivated lamb (or goat) serum in 1× PBS.

10. Primary antibodies: Most common primary antibodies to show cell polarity in *Xenopus* embryos are ZO-1 (Invitrogen #61-7300), Cingulin (Invitrogen #36-4401), P-E-cadherin (Epitomics #2239-1), β-catenin (Santa Cruz #sc-7199), PKC-ζ (Santa Cruz #sc-216), etc.

11. Secondary antibodies: Most common secondary antibodies used in *Xenopus* embryos are Cy3-conjugated donkey anti-rabbit IgG (ImmunoResearch Lab #711-165-152), FITC-conjugated goat anti-mouse IgG (Invitrogen #62-6511), etc.

12. Vectashield with DAPI: Vectorlabs #H-1200.

13. Nail polish.

3. Methods

3.1. Obtaining Embryos

1. Obtain fertilized embryos: Natural mating is the most effective method for obtaining the most symmetrically and orderly cleaving embryos at 32-cell stage. For this method, both male and female frogs are primed by hormone injections. Typically, males receive an injection of 50 IUs of HCG and females receive an injection of 1,000 IUs 16 h before the experiment. Place the male and female frogs in a 30 l tank filled with 16 l of 0.1× MBS 12 h prior to the time when fertilized eggs are desired (see Note 1). The bottom of the tank should contain square petri dishes covered with a stiff plastic screen. The frogs should be left in the dark (we drape the chamber with black cloth) for the next 24 h. As eggs are laid, they drop through the plastic screen into square petri dishes, and can be collected throughout the day. We use a specially constructed acryl chamber for this purpose (Fig. 1).

2. Remove the jelly coats from fertilized eggs that have just begun to cleave by gently swirling the eggs in 4 volumes of dejellying solution (L-cysteine in 0.5× MBS) for 4–5 min. After dejellying, immediately wash embryos five times with 0.5× MBS.

3. Transfer embryos to fresh 3% ficoll/0.5× MBS in a clean petri dish. Store at 13, 18, and 23°C until they reach the 32-cell stage.

3.2. Identifying Specific Blastomeres

If the study requires localization of the marker to specific tissues or regions, it is essential to identify the specific blastomere that will give rise to that tissue. According to the fate maps constructed by Dale and Slack (14) and Moody (9), one can identify a specific blastomere with distinct developmental fates in 32-cell stage embryos. In fertilized eggs, the dorsal side of the embryo can be predicted very accurately (>90%) by noting the orientation of the first cleavage furrow. At fertilization, pigmentation of the animal hemisphere begins to contract towards the sperm entry point on the ventral side, causing the dorsal equatorial region to become less pigmented. If the first cleavage furrow bisects this lighter area equally between the two daughter cells, then that lighter area can be used as the indicator of the dorsal side, and the first cleavage furrow will indicate the midsagittal plane (15, 16). Accordingly, individual blastomeres are designated as D (dorsal side) or V (ventral side) when the embryo cleaves to 4 cells. Numbers are added to the blastomere's nomenclature as the embryo further divides; these numbers indicate the cell's position, which in turn predicts which specific tissue or region it will form. For example, in the cell movement assay in which we found that our protein of interest (i.e., ephrinB1) co-opts the PCP signaling pathway via the scaffold protein Dishevelled to encourage retinal progenitor cells to move into the eye field (12), we targeted the D1.1.1 (also called A1) blastomere, which is a major contributory cell to the retinal field (~50%) later in development. In contrast, another blastomere V1.1.1 (also called A4), which is a non-contributor (<1%) to the retinal field is normally fated to move into and populate head and trunk epidermis. Since ephrinB1 RNA can cause a portion of the V1.1.1 progeny to disperse into the eye field and populate retina, this allowed us to test ephrinB1-driven cell movement and whether PCP signaling was critical for this event, and whether modulators (i.e., FGFR) can regulate this activity (Fig. 2; 11–13).

3.3. Injections into a Specific Blastomere at 32-Cell Stage Embryos

Before starting a specific blastomere injection at 32-cell stage embryos, careful titration of all mRNAs should be done. Two commonly used mRNA tracers are β-galactosidase and Green Fluorescent Protein (GFP). Both proteins are derived from non-vertebrates (β-galactosidase from bacteria and GFP from jellyfish), can be distinguished from endogenous vertebrate proteins, are too large to diffuse through gap junctions and have no known deleterious effects on developing vertebrate cells. However, excessive amounts of GFP,

D1.1.1 (A1)

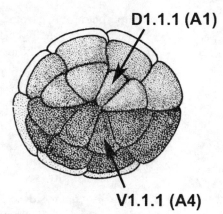

V1.1.1 (A4)

Fig. 2. Experimental scheme for a specific blastomere injection. Animal pole view of a 32-cell embryo. *Lighter blastomeres* depict the dorsal side and *dark blastomeres* depict the ventral side. D1.1.1 is also called A1; V1.1.1 is also called A4. Modified from Nieuwkoop and Faber developmental series (17).

β-galactosidase or the other mRNAs or MOs (morpholino antisense oligonucleotides) can cause non-specific side effects to embryos. Therefore, one needs to carefully test for the optimal concentration of the reagent to be injected. According to our experiments, generally 100–250 pg of GFP or β-galactosidase mRNA provide a strong signal and avoid non-specific side effects. For test mRNAs or MOs, titration should be done before every experiment and injection volumes should be kept to 1–2 nl per blastomere to avoid damaging the cells. For example, to examine the role of canonical and non-canonical Wnt signaling in cell movement, epistasis experiments using dominant-negative TCF3 mRNA (DN-Tcf3; a transcription factor that associates with β-catenin to activate canonical Wnt target genes) and an MO to Daam1 (a protein with Formin homology that binds to Xdsh and activates the small GTPase Rho required by the PCP pathway) can be used. Introduction of DN-Tcf3 mRNA into a particular blastomere at levels that inhibit Wnt3a-induced secondary axes should be used. In our case, this amount is 300 pg, but each mRNA should be titered for an appropriate effect in a two-cell embryo injection. Daam1 MO (5′-GCCGCAGGTCTGT CAGTTGCTTCTA-3′) was introduced at 15 ng and found to cause PCP-associated defects (12).

3.4. Fixation and Histochemical Reactions

3.4.1. Fixation and Histochemical Reactions for β-Galactosidase

1. Harvest embryos at stage 12.5 (17) and put them in glass scintillation vials (see Note 2).

2. Rinse the embryos with 0.7× PBS.

3. Fix the embryos for 1 h at room temperature with freshly made 0.7× PBS containing 2% Formaldehyde, 0.2% Glutaraldehyde, 0.02% Triton X-100, and 0.01% Sodium Deoxycholate (see Note 3).

4. Rinse the embryos twice with 0.7× PBS.

5. Stain the embryos at 30°C in 0.7× PBS containing 5 mM $K_3Fe(CN)_6$, 5 mM $K_4Fe(CN)_6$, and 1 mg/ml X-gal (or Red-gal), 1 mM $MgCl_2$ (see Note 4).

6. When staining is complete, refix the embryos in MEMPFA for 1 h at room temperature.

7. If the color of staining is difficult to distinguish because of the natural pigmentation of embryos, add bleaching solution to eliminate the pigments. Incubate stained and refixed embryos in the bleaching solution for 1 h at room temperature under a fluorescent light (see Note 5).

8. Check β-galactosidase staining under the stereomicroscope, otherwise stained embryos can be stored in the dark at 4°C for 1 day (see Note 6).

3.4.2. Fixation, Embedding, and Sections for GFP

1. Harvest embryos at stage 37–38 (17) and put them in glass scintillation vials (see Note 7).

2. Rinse the embryos with 1× PBS.

3. Fix the embryos for 1 h at room temperature with freshly made 100 ml of 1× PBS containing 4% Paraformaldehyde, 0.9 g of NaCl, 40 ml of 0.2 M of Na_2HPO_4, 10 ml of 0.2 M of NaH_2PO_4 (see Note 8).

4. When embryos are ready for further processing, they should be rinsed three times in 1× PBS for 10 min.

5. Transfer the embedded embryos with small amount of 1× PBS into embedding mold.

6. Remove all residual PBS from embedding mold (see Note 9).

7. Immediately add OCT compound into the embedding mold (see Note 10).

8. Orient the tadpoles vertically with sharp dissecting needle under the microscope (Fig. 3b; see Note 11).

9. Freeze on dry ice with methanol for 15–30 min until OCT compound completely hardens (see Note 12).

10. Remove the block from the embedding mold (see Note 13).

11. Attach the block to the cryostat chuck using small volume of OCT compound (see Note 14).

12. Equilibrate the block at least for 30 min in the cryostat.

13. Cut 16 µm sections at –17°C of object temperature, –25°C of chamber temperature.

14. Gently stretch out the cut sections with brush while on the cryostat so they remain cold, and place glass slide over sections and they will transfer to slide within a second (see Note 15).

15. Dry the slide (laying down) at room temperature for 30 min.

16. Mount the slides with a couple of drops of Vectashield (see Note 16).

17. Apply the cover slips onto the slides (see Note 17).

18. Seal the slides/cover slips with nail polish.

19. Check GFP under the fluorescence microscope, otherwise mounted sections can be stored in the dark at –80°C.

3.4.3. Fixation, Embedding, Sections, and Antibodies Staining for Immunofluorescence

1. Harvest embryos at stage 10.5, 15, or 37–38 (17) and put them in glass scintillation vials (see Note 18).

2. Remove all the 0.1× MBS.

3. Fix embryos in cold Dent's fixative (4 volumes of methanol, 1 volume of DMSO) overnight at –20°C or for 2 h at room temperature (see Note 19). After fixing the embryos, fixed embryos should be placed at –20°C until the next steps are ready.

4. When embryos are ready for further processing, they should be rinsed three times in 1× PBS for each 10 min.

5. Remove all the residual 1× PBS from glass scintillation vials.

6. Embed embryos in 15% cold fish gelatin/sucrose solution for 24 h at 4 °C or for 2 h at room temperature. The embedded samples can be stored for up to 2 weeks at 4°C when 0.02% Sodium azide added (see Note 20).

7. Pour the fresh 15% cold fish gelatin/sucrose solution into embedding mold.

8. Transfer the embedded embryos into embedding mold.

9. Orient the embryos vertically with sharp dissecting needle under the microscope (Fig. 3a; see Note 21).

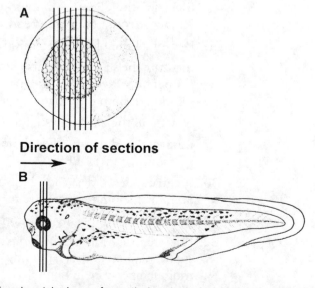

Fig. 3. Experimental schemes for sectioning. *Arrow* depicts direction of sections at early embryo ((**a**); stage 10.5) and tadpole ((**b**); stage 37–38). Modified from Nieuwkoop and Faber developmental series (17).

10. Freeze on dry ice with methanol for 15–30 min until the gelatin/sucrose solution completely hardens (see Note 22).

11. Remove the block from the embedding mold (see Note 23).

12. Attach the block to the cryostat chuck using small volume of OCT compound (see Note 24).

13. Equilibrate the block at least for 30 min in the cryostat.

14. Cut 10 μm sections at –17°C of object temperature, –25°C of chamber temperature.

15. After making five cut sections, the sections should be immediately transferred to polylysine-coated glass slides (see Note 25).

16. Dry the slide at room temperature for 30 min in the fume hood.

17. Store slides at –70°C until they are ready for further processes.

18. Dry the slides at room temperature for 1 h in the fume hood.

19. Dip the slides in acetone for 1 min (see Note 26).

20. Dry the slides at room temperature for 10 min.

21. Draw a boundary line around the slide edges using the PAP pen (see Note 27).

22. Transfer the slides to a reservoir of 1× PBS and quickly rehydrate the slides.

23. Incubate the slides horizontally with Image-iT signal enhancer for 30 min at room temperature in a humidity chamber (Fig. 4; see Note 28).

24. Wash the slides three times with 1× PBS.

25. Incubate the slides with blocking solution horizontally for 30 min at room temperature in a humidity chamber.

26. Decant all the blocking solution from the slides.

27. Incubate the slides horizontally overnight at 4°C or 2 h at room temperature, with the primary antibody diluted in blocking solution in humidity chamber (see Notes 29 and 30).

28. Decant all the antibody solution from the slides.

29. Return the slides to slide holder and wash three times for 10 min each in 1× PBS.

30. Incubate the slides with blocking solution horizontally for 30 min at room temperature in a humidity chamber.

31. Decant all the blocking solution from the slides.

32. Place the slides in the humidity chamber and cover each with appropriate secondary antibody and incubate slides for 2 h at room temperature in humidity chamber (see Notes 31 and 32).

33. Return the slides to slide holder and wash three times for 15 min with 1× PBS at room temperature.

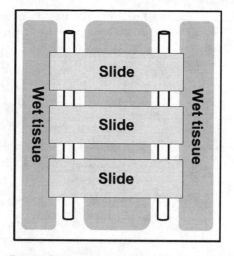

Fig. 4. Schematic diagram for humidity chamber. Cartoon depicting top view of the humidity chamber needed for antibody staining.

34. Dry the slides at room temperature for 1 min.

35. Mount the slides with a couple of drops of Vectashield with DAPI (see Note 33).

36. Apply the cover slips onto the slides.

37. Seal the slides/cover slips with nail polish.

38. Check the immunofluorescence under the fluorescence microscope, otherwise mounted sections can be stored in the dark at −80°C.

4. Notes

1. Alternatively, 10% Steinberg's or 0.1× MMR solution can be used for natural mating.

2. At stage 12.5, D1.1.1 clones are broadly dispersed across the dorsal animal quadrant, which includes the retinal field, whereas V1.1.1 clones remain on the ventral side (18). Changes in these positions can be detected by the lineage tracers, and thereby the effects of the test mRNA on the early movements of retinal progenitor cells during gastrulation can be monitored.

3. Alternatively, embryos can be fixed at 4°C overnight.

4. Staining time can vary from 30 min to 2 h. Do not exceed 2 h, since it causes too strong background staining.

5. Bleaching time can vary from 1 to 4 h. Do not exceed 4 h, since it can damage the embryos. If embryos need more bleaching, transfer them to new bleaching solution and bleach only for 10 min more.

6. For best results, capture images of the embryos immediately because the signal may decrease very rapidly after 1 day.

7. At stage 37–38, D1.1.1 clones are nicely localized to the retinal field, whereas V1.1.1 clones are confined to the head and trunk epidermis.

8. Alternatively, embryos can be fixed at 4°C overnight.

9. All liquid must be removed completely before adding OCT compound, otherwise any residual buffer may cause thawing of the block when the block is touched by hand.

10. If embryos are exposed to air, they may be damaged.

11. When embryos are sectioned horizontally, the sections may not be symmetrical between left and right sides.

12. Care must be taken when embedding mold is on dry ice/methanol, since this system is extremely cold.

13. This is easier when the block is still cold and very hard. There is no need to trim the block. Keep it frozen on dry ice.

14. Put a couple of drops of liquid OCT compound on the chuck, place the still frozen block onto it and then transfer immediately to the cryostat.

15. Waiting too long to transfer the section may result in the section rolling up.

16. Vectashield can reduce the background fluorescence as well as allow clear sample fluorescence.

17. Care must be taken when applying the cover slip as the frozen sections are very fragile. Avoid air bubbles and do not press the cover slip too hard, as both actions will smash the sections.

18. At appropriate stages, we can check the localization of cell polarity-related proteins, such as aPKC and E-cadherin. The time at which the embryos should be harvested is based on the purpose of each experiment.

19. If fixing at –20°C, all liquids must be removed completely by performing a couple of washes with cold Dent's fixative before placing at –20°C, otherwise any residual water may freeze.

20. Sodium azide has to be added only if the fish gelatin is stored for a long time.

21. This is much easier when the gelatin/sucrose solution is cooled on ice, as it becomes more viscous.

22. Once frozen, the samples must be sectioned on the same day, ideally within 4–5 h.

23. This is easier when the block is still cold and very hard. There is no need to trim the block. Keep it frozen on dry ice.

24. Put a couple of drops of liquid OCT compound on the chuck, place the still-frozen block onto it and then transfer immediately to the cryostat.

25. Waiting too long to transfer the section may result in the section rolling up.

26. Place the slides in a slide holder and place it in a reservoir containing acetone for 1 min. Do not exceed 1 min.

27. This stops liquid running off the edges of the slides.

28. Image-iT signal enhancer can reduce the background fluorescence as well as allow clear sample fluorescence.

29. Alternatively, the slides can be incubated for 2 h at room temperature.

30. The individual slide is covered with 0.2 ml of diluted antibody solution. The dilution factor for each antibody is different. We typically use a 1:200 dilution.

31. Add 0.2 ml of appropriate secondary antibodies to cover the sections. The dilution factor for each antibody is different. We typically use a 1:400 dilution. Do not apply secondary antibodies at a concentration greater than 1:400 since it causes high background fluorescence.

32. This and all subsequent steps must be performed in the dark.

33. Care must be taken when applying the cover slip, as the frozen sections are very fragile. Avoid air bubbles and do not press the cover slip too hard, as both actions will smash the sections.

References

1. Simons, M., Mlodzik, M. (2008) Planar cell polarity signaling: from fly development to human disease. *Annu Rev Genet* 42, 517–540.

2. Vladar, E.K., Antic, D., Axelrod, J.D. (2009) Planar cell polarity signaling: the developing cell's compass. *Cold Spring Harb Perspect Biol* 1, a002964.

3. Lee, M., Vasioukhin, V. (2008) Cell polarity and cancer--cell and tissue polarity as a non-canonical tumor suppressor. *J Cell Sci* 121, 1141–1150.

4. Guo, P., Weinbaum, A.M., and Weinstein, S. (2003) A dual-pathway ultrastructural model for the tight junction of rat proximal tubule epithelium. *Am J Physiol Renal Physiol* 285, 241–257.

5. Handler, J.S. (1989) Overview of epithelial polarity. *Annu Rev Physiol* 51, 729–740.

6. Rodriguez-Boulan, E., Powell, S.K. (1992) Polarity of epithelial and neuronal cells. *Annu Rev Cell Biol* 8, 395–427.

7. Fleming, T.P., McConnell, J., Johnson, M.H., Stevenson, B.R. (1989) Development of tight junctions de novo in the mouse early embryo: control of assembly of the tight junction-specific protein, ZO-1. *J Cell Biol* 108, 1407–1418.

8. Eckert, J.J., Fleming, T.P. (2008) Tight junction biogenesis during early development. *Biochim Biophys Acta* 1778, 717–728.

9. Moody, S.A. (1987) Fates of the blastomeres of the 32-cell-stage *Xenopus* embryo. *Dev Biol* 122, 300–319.

10. Lee, H.S., Nishanian, T.G., Mood, K., Bong, Y.S., Daar, I.O. (2008) EphrinB1 controls cell-cell junctions through the Par polarity complex. *Nat Cell Biol* 8, 979–986.

11. Moore, K.B., Mood, K., Daar, I.O., Moody, S.A. (2004) Morphogenetic movements underlying eye field formation require interactions between the FGF and ephrinB1 signaling pathways. *Dev Cell* 6, 55–67.

12. Lee, H.S., Bong, Y.S., Moore, K.B., Soria, K., Moody, S.A., Daar, I.O. (2006) Dishevelled mediates ephrinB1 signalling in the eye field

through the planar cell polarity pathway. *Nat Cell Biol* 8, 55–63.

13. Lee, H.S., Mood, K., Battu, G., Ji, Y.J., Singh, A., Daar, I.O. (2009) Fibroblast growth factor receptor-induced phosphorylation of ephrinB1 modulates its interaction with Dishevelled. *Mol Biol Cell* 20, 124–33.

14. Dale, L. and Slack, J.M.W. (1987) Fate map for the 32-cell stage of *Xenopus laevis*. *Development* 99, 527–551.

15. Klein, S.L. (1987) The first cleavage furrow demarcates the dorsal-ventral axis in *Xenopus* embryos. *Dev Biol* 120, 299–304.

16. Masho, R. (1990) Close correlation between the first cleavage plane and the body axis in early *Xenopus* embryos. *Dev Growth Diff* 32, 57–64.

17. Nieuwkoop, P.D. and Faber, J. (1967) Normal Table of *Xenopus laevis*, 2nd Ed. North-Holland, Amsterdam.

18. Bauer, D.V., Huang, S. and Moody, S.A. (1994). The cleavage stage origin of Spemann's Organizer: Analysis of the movements of blastomere clones before and during gastrulation in *Xenopus*. *Development* 120, 1179–1189.

Chapter 9

Gene Loss-of-Function and Live Imaging in Chick Embryos

Anne C. Rios, Christophe Marcelle, and Olivier Serralbo

Abstract

Planar cell polarity (PCP) is the coordinate organization of cells within the plane of a tissue. PCP is essential for tissue function, such as for proper hearing in the vertebrate ear or for accurate vision in the Drosophila eye. Using the chick embryo, we have recently shown that during early muscle formation, the first formed muscle fibres utilize the PCP pathway to orient parallel to a WNT11 source present in the medial border of somites. Our results further establish that WNT11 acts as a directional cue to regulate this process. To perform this study, two major techniques have been utilized, the gene loss-of-function using a vector-based shRNAmir expression and confocal videomicroscopy of fluorescent gene reporters targeted in specific cell subpopulations by in vivo electroporation. Here we describe the two techniques.

Key words: Chick embryo, Somite, shRNAmir, Confocal videomicroscopy, In vivo imaging

1. Introduction

Because of its easy accessibility, the chick embryo has been instrumental to numerous seminal discoveries in the field of developmental biology. However, its usefulness has been hampered by the difficulty to perform genetic manipulations. Novel technologies developed in the past years have opened new routes of investigation. The development of the electroporation technique (1, 2) allows the targeted expression of any vector in specific cell subpopulations. Fluorescent gene reporters have been utilized for analyses of cell and tissue movements in normal conditions or after gain or loss of gene function. After the discovery of successful induction of RNA interference in mammalian cells, vector-based siRNA expression systems have been developed, in which a short-hairpin RNA (shRNA) is expressed by RNA polymerase III promoter, such as U6 or H1 (3–5) although RNA polymerase II promoter has also been utilized (6). The use of shRNAmir has the advantage to

Kursad Turksen (ed.), *Planar Cell Polarity: Methods and Protocols*, Methods in Molecular Biology, vol. 839,
DOI 10.1007/978-1-61779-510-7_9, © Springer Science+Business Media, LLC 2012

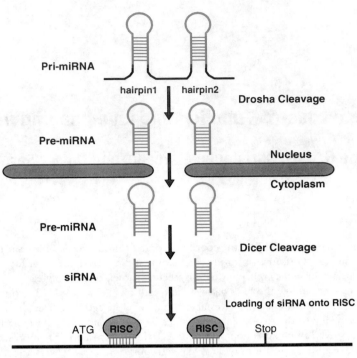

Fig. 1. Maturation process of the pri-miRNA into two siRNAs. Pri-miRNA, which contains the two hairpins, is produced under the control of the U6 promoter. The digestion by Drosha enzyme gives rise to two Pre-miRNAs, which are exported outside the nucleus by the enzyme exportin into the cytoplasm. The pre-miRNAs give rise to siRNAs after digestion by Dicer and onto RISC for target recognition and RNA degradation.

reduce the interferon response and to increase the efficiency of the response (7) by mimicking endogenous microRNAs, processed by the siRNA maturation pathway via Dicer and RISC (Fig. 1). A vector developed by Das et al. (8) uses a derivative of human mir30 driven by a chicken U6 promoter, shown to be more efficient in chick tissues than the mammalian version (8). Compared to alternative techniques to downregulate gene expression in somite (morpholino, shRNA), we found that this approach is the most efficient.

Cell and tissue movements are crucial for embryonic development but they are difficult to evaluate in fixed samples. Exquisite details of the dynamic cell shape changes taking place during embryogenesis are revealed by time-lapse confocal in vivo imaging.

2. Materials

2.1. ShRNAmir Construct

1. pRFPRNAiC cloning vector for the overexpression of two shRNAmir (Das et al. (8), ARK genomics).

2. Synthesized oligos for gene-specific shRNAmir expression sequence. The design of target sequences was performed by

submitting the coding sequence of the target gene to GenScript (https://www.genscript.com/ssl-bin/app/rnai/). Target sequences must be 22 nucleotides long to keep the structure of miRNA correct. Oligos (Sigma) were purified by HPLC.

3. DNA cloning: *NheI, MluI, SphI,* T4 DNA ligase (New England Biolabs), TSAP (Thermosensitive Alkaline Phosphatase, Promega), Phusion high-fidelity Taq polymerase (Finnzymes).

4. DNA purification: Wizard Plus SV minipreps and Wizard SV Gel and PCR clean-up kits (Promega). NucleoBond Maxiprep Kit PC500 (Macherey-Nalgen).

5. LB Broth.

6. LB agar for LB plate. After melting and cooling down to 50°C, add ampicillin (final concentration 100 µg/ml) and pour into 10 cm Petri dishes. Store at 4°C.

7. Ampicillin stock solution 100 mg/ml stored at –20°C.

8. E. coli chemically competent cells One Shot Top10 (Invitrogen).

9. UltraPure agarose (Invitrogen).

10. 1-kb and 100-bp DNA ladder (Promega), stored at 4°C.

2.2. Somites Electroporation

1. Fertilized eggs are ordered from a local poultry farm and maintained at 16°C before incubation at 38°C for 60 h or stage 14 HH (9) of embryonic development.

2. Howard Ringer's solution stock prepared 10× and sterilized.

3. Penicillin/streptomycin Sigma 100× aliquots in 2 ml tubes and stored at –20°C. 2 ml are diluted in 50 ml Ringer 1× solution.

4. Electroporation mix is prepared as described in ref. 10.

5. Indian ink (Lefranc Bourgeois Nan King) sterilized and aliquoted in 2 ml tube, store at 4°C.

6. Electroporator intracell TSS20 ovodyne and EP21 current amplifier.

7. Electrodes are made with platinum wire 5/10 mm diameter for the positive and tungsten wire 0.375 mm diameter for the negative. Embryos are protected from electrical burn by insulating the wire with nail polish as described in ref. 10.

8. Capillaries from Harvard Appartus model GC120T-10.

9. Pipette puller Sutter P-2000.

10. Stereo microscope Zeiss Lumar V12 equipped with camera Zeiss HRc.

11. Confocal Zeiss LSM510.

2.3. In Situ Hybridization

1. PBTw: PBS, Tween20 0.1%.

2. Glutaraldehyde 25% high grade (Sigma).

3. Proteinase K (Sigma) is dissolved in water at 20 μg/μl.

4. Hybridization Mix: Formamide 50%, SSC 1.3× (pH5 adjusted with citric acid), Yeast RNA 50 μg/ml, Tween-20 0.2%, CHAPS 0.5%, Heparin 100 μg/ml.

5. MABT: 100 mM Maleic Acid, 150 mM NaCl, 0.1% Tween-20, final pH 7.5.

6. Anti-Digoxigenin alkaline phosphatase antibody (Roche).

7. Boehringer Blocking Reagent (BBR) prepared in MABT solution.

8. HyClone Characterized Foetal Bovine Serum (FBS) from Thermo Scientific.

9. NBT/BCIP (Promega).

10. NTMT: NaCl 100 mM; Tris–HCl pH9.5 100 mM; $MgCl_2$ 50 mM; Tween20 0.1%.

2.4. Antibody Staining

1. Rabbit Anti-RFP polyclonal antibody (AbCam).

2. Goat Anti-Rabbit 555 Alexa Fluor secondary antibody (Invitrogen).

3. PBS, 0.1% BSA, 0.1% triton X-100, 0.2% SDS.

2.5. Whole Mount Embryo Preparation

1. Glycerol 50% in PBS.

2. Glycerol 80% in PBS.

2.6. Time-Lapse Experiments

1. UltraPure™ Low Melting Point Agarose (Invitrogen).

2. F-12 Nutrient Mixture (1×), liquid (contained Glutamax) from invitrogen.

3. Foetal Bovine Serum (FBS) from Thermo Scientific.

4. Sodium pyruvate from invitrogen.

5. Penicillin/streptomycin 100× from Sigma.

6. Glass bottom dish from WillCo-Dish (ø 12 mm).

3. Methods

To study the role of PCP signaling in myocyte polarity, we have downregulated gene expression using the shRNAmir technology and performed time-lapse experiments to visualize myocyte elongation. To illustrate these two methods, we present the shRNAmir-mediated dowregulation of the secreted protein WNT11. The role of WNT11 as a positional cue for myocyte elongation has been shown in (11). Downregulation of WNT11 expression in somite resulted in polarity defects of myocytes (Fig. 2b). Live confocal video microscopy of somite cells electroporated with GFP and

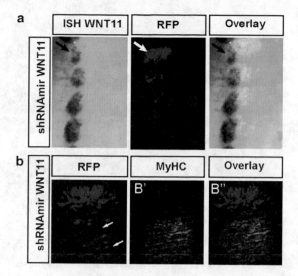

Fig. 2. *Wnt11* downregulation and PCP phenotypes. (**a**) *Wnt11* in situ hybridization, *Wnt11* expression is downregulated in the medial border of electroporated somites (*arrow*). (A′) Expression of the shRNAmir in electroporated cells visualized by the RFP reporter. (A″) overlay showing the electroporated somites where *Wnt11* mRNA is downregulated. (**b**) Whole mount confocal imaging of RFP electroporated cells showing defects in growth and polarization (*arrows*) of the differentiated myocytes visualized by Myosin Heavy Chain (MyHC) immuno-staining (B′). (B″) overlay.

Histone2B-RFP (H2BRFP) reporter genes shows intense plasma membrane protruding activity of epithelial somite cells, cell division and the initiation of myocyte elongation (Fig. 3f–g).

3.1. Plasmid Construct

To downregulate gene expression in chick somites, we have found that the most reliable method is the vector-based shRNAmir technique. The cloning vector contains a fluorescent marker, red fluorescent protein (RFP), that allows the detection of electroporated cells. This is important, since electroporation results in a mosaic transfection of the targeted cell population. The shRNAmirs, driven by a chick U6 promoter, are continuously produced, thereby compensating the siRNA degradation and the dilution resulting from cell divisions. The sequence targets are chosen by submitting the gene coding sequence (in this case chick *Wnt11*) to GenScript target finder. The targets are 22 nucleotides long with a 30–60% GC ratio. BLAST of the sequences against the chick genome should return little similarity except for the target gene. Four oligonucleotides are necessary to design each shRNAmir (Fig. 4a, b): two universal primers contain the cloning sites and overlap with specific primers directed against the target sequence (target primers). The 5′ base target sequence of the sense strand has been changed to mimic the miRNA30 mismatches (Fig. 4c). The PCR reaction using these four primers as matrice gives rise to a DNA fragment containing the cloning sites and the sequence of the shRNAmir.

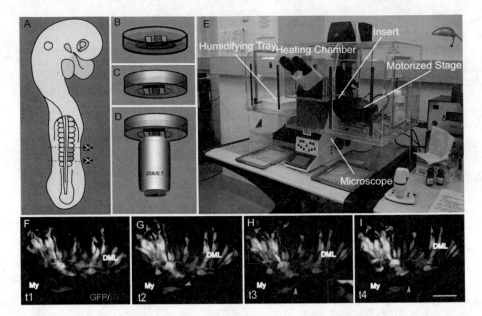

Fig. 3. Live imaging of chick embryos. (**a–d**) Schemes showing (**a**) a 3 day-old chick embryo, with three somites electroporated on the right side. The *red dashed lines* surround the truncal part of the embryo, which were dissected and prepared
for imaging. (**b**) The sample is placed in a glass-bottom dish to equilibrate in F12 glutamax medium. (**c**) The dorsal part of the
sample is placed against the glass. (**d**) The electroprated somites are observed with an inverted confocal microscope with
a 20× objective. (**e**) The confocal microscope used for the experiments in its heating chamber. (**f–g**) Snapshots of a movie
(see Supp. Movie1) showing a dorsal view of somite cells co-electroporated with pCAGGS-EGFP (in *green*) and pCAGGS-
H2B-RFP (in *red*). *Red arrowheads* show protrusions sent out by epithelial cells. *Yellow arrowhead* shows a cell in the
myotome that initiates its elongation in the antero-posterior axis. The purple box shows a magnified view of a cell in division. Abbreviations: *DML* dorso-medial lip, *TZ* transition zone, *My* Myotome. Scale bars: 50 μm.

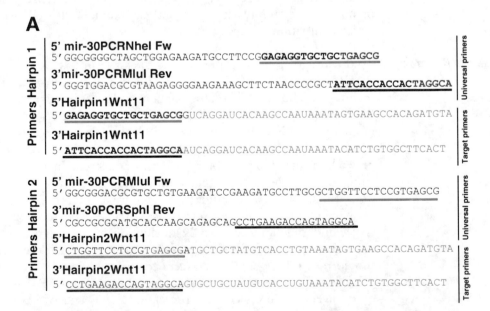

Fig. 4. ShRNAmir construct. (**a**) Primers sequences of the two ShRNAmir directed against *Wnt11*. Each Hairpin is composed
of two target primers containing the specific target sequences (*blue*) and the loop (*green*), and two universal primers containing
the cloning sites and flanking sequences. Overlapping sequences are underlined with colour code. (**b**) Overlapping organization
of the primers and cloning process for a vector-based shRNAmir. (**c**) Resulting ShRNAmir from Hairpin1 primer with corresponding
colour codes. Digestion sites by maturation enzymes Drosha and Dicer are shown in *black* and *grey* respectively.

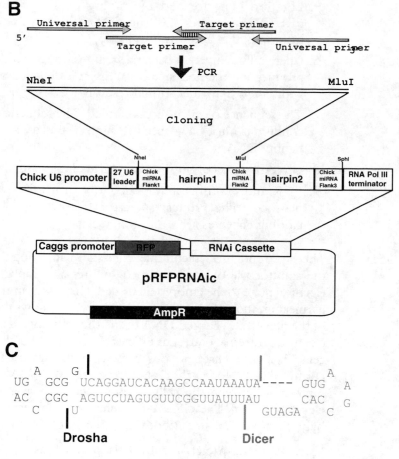

Fig. 4. (continued)

Two shRNAmir can be introduced in the cloning vector. We routinely use two shRNAmir for each target gene to increase the efficiency of the gene loss of function.

1. HPLC-purified primers are ordered at Sigma.

2. The primers are diluted at 1 μg/μl for stock solution. For PCR, 100 ng for the universal primer and 10 ng of specific primer are mixed with Phusion high-fidelity Taq polymerase in a PCR volume of 50 μl. Annealing 55°C 30′, extension 72°C 30′, denaturation 94°C 30′ during 35 cycles.

3. The PCR is analyzed by electrophoresis on 2% agarose gels, a 170-bp band is expected.

4. PCR products are purified using Wizard PCR cleanup and digested by NheI and MluI. The pRFP-RNAiC vector is also digested by NheI/MluI in the presence of TSAP to dephosphorylate the 5′ extremities. TSAP is inactivated at 74°C, 15 mn.

5. Ligation is performed by mixing 50 ng of the digested pRFPRNAiC vector, 200 ng of the digested PCR fragment, ligation buffer, and T4DNA ligase, O/N at 16°C.

6. Chemically competent bacteria are transformed with the ligation product and the presence of an insert is detected by digestion with NheI/SphI that should result in a 200-bp fragment.

7. The sequence of the insert is verified. The sequencing reaction conditions must be adapted to high secondary structure-containing DNA.

8. Glycerol stocks of positive bacterial colonies are made.

9. PCR and cloning of the second hairpin follow the same protocol. The primers set are different (Fig. 4a) and the cloning enzymes are *MluI* and *SphI*.

3.2. Electroporation

The electroporation technique consists in injecting DNA inside the somitocoele of newly formed somites and applying a square electrical pulse to the embryonic tissues. This results in the transitory permeabilization of plasmic membrane, which allows the migration of the negatively charged DNA inside the cells towards the positive electrode. Electroporation is efficient at the apical end of epithelial cells. The adherens junctions present at the apico-lateral end of epithelial cells force the DNA to enter the cells rather than migrating between the cells. Mesenchymes are not receptive to electroporation. The siRNA efficiency can be tested by in situ hybridization or, if an antibody is available, by immuno-staining.

1. 100 ml of bacterial culture is purified using NucleoBond PC500 Maxiprep Kit, the DNA pellet is resuspended in 50 µl of water. This results in a highly concentrated DNA (between 5 and 10 µg/µl). The electroporation solution contains 2 µg/µl final concentration of DNA together with 5 µl of concentrated electroporation mix in a final volume of 15 µl.

2. Fertilized eggs are maintained at 16°C and incubated for various periods at 38°C in humid atmosphere. For somite electroporation, eggs are incubated for 60 h.

3. To lower the embryo in the shell, 3 ml of albumen is removed with a syringe and the eggshell is windowed with scissors.

4. Indian ink is injected below the embryo with a needle and 300 µl of Howard Ringer's penicillin/streptomycin solution are added onto the embryo.

5. The vitelline membrane covering the embryo is removed with a needle.

6. The DNA mix is aspirated with a rubber tube fitted onto a pulled capillary. The thin tip of the capillary is introduced through the presomitic mesoderm into the somites and the DNA is blown into the somitocoele.

7. The two electrodes are quickly positioned onto each side of the embryo and electric pulses are applied with a foot switch. For electroporation of the medial border of the somite, we use the following settings: 70 V, 10 ms, one pulse train. With the foot switch, apply two or three pulses depending of the age of the embryos, younger embryo being more sensitive to electrical current than old ones.

8. Place a piece of tape to close the eggshells and put them back in the incubator overnight.

9. The presence of RFP is examined under fluorescence stereo microscope. Embryos that display fluorescence at the right place are kept. Burned embryos (i.e., embryos that did not develop normally, compared to controls) are systematically discarded, since wounds lead to artifactual results. On average, 30% of electroporated embryos are kept for further analyses.

10. Embryos are dissected out in PBS, fixed 1 h in 4% formaldehyde in PBS at room temperature or O/N at 4°C.

11. Wash in 50% EtOH in PBS, then in EtOH 100%. Embryos can be kept for months in EtOH 100% at –20°C.

3.3. Whole Mount In Situ Staining

1. Rehydrate embryos through 70, 50, 25% MeOH/PBTw series, wash twice in PBTw.

2. Treat with 20 µg/ml proteinaseK in PBTw. Treat embryos as short as possible with the enzyme. For embryos at 84 h development –15 mn at room temperature.

3. Remove the ProteinaseK solution, rinse briefly, and post-fix for 20 mn in 4% Formaldehyde +0.1% Glutaraldehyde in PBStw.

4. Rinse and wash with PBSTw and transfer in 2 ml round-bottom Eppendorf tube.

5. Rinse with 1:1 PBSTw/Hybridization mix. Let the embryos settle.

6. Rinse with 1 ml hybridization mix. Let embryos settle.

7. Replace with 1 ml hybridization mix and incubate horizontally for 1 h at 65°C.

8. Replace with 1 ml hybridization mix +1/10th of a transcription reaction made with 1 µg of plasmid. Place immediately at 65°C.

9. Incubate O/N at 65°C.

10. Rinse twice with pre-warmed 65°C hybridization mix.

11. Wash two times 30 mn at 65°C with 1 ml of pre-warmed hybridization mix

12. Wash 20 mn at 65°C with 1 ml 1:1 MABT/Hybridization mix and rinse two times with 1 ml MABT.

13. Incubate minimum 1 h in 1 ml MABT+2% BBR+20% FBS.

14. Incubate 6 h at room temperature in 1 ml MABT+2% BBR+20% FBS+Anti-Dig Alkaline phosphatase antibody 1/2,000.

15. Transfer to 15 ml tube and rinse two times with MABT.

16. Wash O/N at room temperature in 10 ml MABT.

17. Wash three times 1 h in MABT.

18. Wash two times 10 mn with NTMT.

19. Transfer embryos in 12 wells plate and incubate with 2 ml NTMT+NBT/BCIP at room temperature. Staining can develop faster at 37°C.

20. If background staining is coming up too fast, stop the reaction by rinsing embryos in MABT overnight at 4°C, repeat steps 19 and 20 as many times as desired for a satisfactory staining. With difficult probes, this procedure can take up to a week for the desired result.

3.4. Whole Mount Antibody Staining

RFP (or GFP) fluorescence is decreasing considerably after in situ hybridization due to the proteinase K treatment. To allow a better visualization of the siRNA electroporated cells, we routinely perform immuno-staining reaction to detect the fluorochrome.

1. Transfer embryos in 2 ml round-bottom tubes.

2. Incubate 1 h embryos in 1 ml MABT+10% FBS at room temperature.

3. Incubate O/N at 4°C with 1 ml of MABT+10% FBS+1/500 anti RFP rabbit polyclonal antibody.

4. Transfer to 15 ml tubes, wash three times for 1 h at room temperature with 10 ml MABT.

5. Transfer to 2 ml tubes and incubate for 4 h at room temperature with 1 ml MABT+10% FBS+1/500 anti-rabbit Alexa Fluor 555 antibody.

6. Transfer to 15 ml tubes, wash in MABT.

7. Pictures can be taken with a dissecting microscope equipped for fluorescence and a colour camera (Fig. 2a).

3.5. Confocal Imaging

1. After whole-mount antibody staining, embryos are washed in 50 and 80% Glycerol/PBS solution. Let embryos settle after each wash.

2. In 80% glycerol, the region of interest is carefully dissected out.

3. Put 3–4 layers of tape on a slide, with a scalpel cut a square in middle to create a chamber.

4. Transfer the dissected tissue in the chamber, position with the dorsal portion of somites facing the top.

5. Place coverslip, the glycerol must fill the whole chamber.

6. Pictures are taken on an inverted confocal microscope Leica SP5. Stacks are taken through the entire thickness of the somite.

3.6. Time-Lapse Experiment

1. Preparation of the inclusion medium
   Prepare 15 ml of F12 Glutamax medium with 0.3 g of LMP agarose. Heat up to 42°C in a beaker with a stirring bar on a stirring heating plate until agarose has melted. Remove from the heating plate, add 5 ml of FCS, 200 µl of NaPyruvate from a stock solution at 100 mM and 500 µl of Pen-Strep 100×.

2. Embryo preparation
   Chick Embryos are co-electroporated with two plasmids, pCAGGS-GFP and pCAGGS-H2B-RFP, at low concentration (0.5 µg/µl). Dissect embryos in PBS1× and remove all the extraembryonic membranes but keeping the embryo intact. Cut, with dissection scissors, a transversal slice of the embryo in the electroporated region. Keep the neural tube and three somites on both sides of the somite of interest.

3. Slice mounting
   Transfer the slice to a glass-bottom Dish. Remove PBS. Add 500 µl of F12 glutamax medium for 5 mn. Remove the F12 medium and add a few drops of F12 agarose medium to the slice. Place the slice bottom up on the glass with the dorsal part of the electroporated somites facing the glass. The tissue should touch the glass. Put the dish on ice to increase the speed of agarose solidification. Add carefully 1 ml of F12 agarose to cover the preparation. Incubate the preparation 1 h at 37°C in a humidified incubator.

4. Confocal or multi-photon live Imaging
   The Inverted SP5 Leica confocal microscope and the Leica multiphoton microscope are equipped with a temperature-controlled chamber (Fig. 3e). Two hours before imaging, turn-on the heating chamber and add water to a humidifying dish in the chamber. Acquisitions are done at 37°C in saturating humidity. The petri-dish is placed on a microscope stage that maintains the dish for scanning. Start the time-lapse experiment overnight.

5. Confocal and multiphoton settings (see Note 1)
   Time-lapse on live tissues is performed during 8 h with one acquisition every 10 min (12). The Z-stack size depends of the sample's thickness (approximately 80 µm). The Z-step size is optimized for 3D reconstruction (0.63 µm). We use 20× multi-immersion (0.70 N.A.) objective with immersion ring setting of 0.17 (with coverglass, water immersion). We use for confocal acquisition, the Argon laser (wavelength 488 nm) and the Diode Pump Solide State (DPSS) 561 nm at laser power 20% (for less toxicity, and because of the strong signals obtained after electroporation). We used the resonant scanner at a frequency of 8,000 Hz with no averaging (500 frames/sec) and we perform one scan with two internal PMT (photomultiplier tubes) activated to image two fluorophores simultaneously

(EGFP and RFP). For multi-photon imaging, the Mai Tai laser (that combines Millennia diode-pumped laser with Tsunami Ti:sapphire technology with a pulse width of 70 femtosecond with 2.5 W of average power) is used with a laser power at 7% at an excitation of 870 nm. We also use two internal PMT to image two simultaneous fluorophores.

4. Notes

1. For time-lapse, the two majors problems encountered in long-term confocal imaging of live sample are the photo-toxicity and the photobleaching. To alleviate these problems we have utilized three technologies.

   The spinning-disk (Nipkow disk) confocal microscope is a multi-beam scanning microscopy well suited for in vivo imaging. Under certain conditions, performances are better than a single beam confocal microscope. Its advantages are the high frame rate, high detection efficiency and due to the low intensity of the excitation light, phototoxicity and photobleaching are minimal. However, sequential acquisition (for two colour imaging) considerably compromises the speed of acquisition and generates alignment problems for co-localisation studies. Moreover, this technology does not allow deep imaging in thick sample (more than 100 µm depth).

   Another alternative is the resonant scanner confocal microscope (such as the ones that equips the SP5 Leica confocal microscope, or the Nikon A1R confocal microscope). It allows a faster scanning of each optical section (up to ten times faster than a classical scanning). The resonant scanning system allows a considerable improvement of time resolution to investigate dynamic events. The higher frame rate and the minimal time of scanning reduce considerably the effects of phototoxicity. The drawback of this technology is the need of intensely fluorescent samples to obtain a good ratio signal/noise.

   The best alternative for thick tissues and lower signals is the Multiphoton imaging. Multiphoton excitation reduced photobleaching and damage to biological samples. This is due to a localized excitation at a focal plane and not throughout the entire sample as in classical confocal systems. Importantly, this technology allows deep imaging within thick tissues. The focal absorption allows a better three-dimensional reconstruction. Nowadays, microscopes exist that have two photons imaging and resonant scanning, theoretically combining the advantages of multiphoton technology and high-speed rate.

Acknowledgments

This study was funded by the Agence nationale de la recherche (ANR) and the EU sixth Framework Programme Network of Excellence MYORES. Martin Scaal and Jerome Gros have initiated the in vivo video microscopy. The help of Pascal Weber, Stephen Firth, and Chad Johnson from Imaging Facilities (IBDML, Marseille and MMI, Monash University) is acknowledged.

References

1. Momose T, Tonegawa A, Takeuchi J, Ogawa H, Umesono K, Yasuda K. (1999). Efficient targeting of gene expression in chick embryos by microelectroporation. *Dev. Growth Differ.* **41**, 335–44.

2. Muramatsu T, Mizutani Y, Ohmori Y, Okumura J. (1997) Comparison of three nonviral transfection methods for foreign gene expression in early chicken embryos in ovo. *Biochem. Biophys. Res. Commun.* **230**, 376–80.

3. Yu, J.Y., DeRuiter, S.L. and Turner, D.L. (2002) RNA interference by expression of short-interfering RNAs and hairpin RNAs in mammalian cells. *Proc. Natl. Acad. Sci. U.S.A.* **99**, 6047–6052.

4. Brummelkamp, T.R., Bernards, R. and Agami, R. (2002) A system for stable expression of short interfering RNAs in mammalian cells. *Science* **296**, 550–553.

5. Paddison, P.J., Caudy, A.A., Bernstein, E., Hannon, G.J. and Conklin, D.S. (2002) Short hairpin RNAs (shRNAs) induce sequencespecific silencing in mammalian cells. *Genes Dev.* **16**, 948–958.

6. Zhou, H., Xia, X.G. and Xu, Z. (2005) An RNA polymerase II construct synthesizes short-hairpin RNA with a quantitative indicator and mediates highly efficient RNAi. *Nucleic Acids Research.* **33**, e62.

7. Silva, J.M. et al. (2005) Second-generation shRNA libraries covering the mouse and human genomes. (2005) Second-generation shRNA libraries covering the mouse and human genomes. *Nat. Genet.* **37**, 1281–1288

8. Das RM, Van Hateren NJ, Howell GR, Farrell ER, Bangs FK, Porteous VC, Manning EM, McGrew MJ, Ohyama K, Sacco MA, Halley PA, Sang HM, Storey KG, Placzek M, Tickle C, Nair VK, Wilson SA. (2006). A robust system for RNA interference in the chicken using a modified microRNA operon. *Dev Biol.* **294**, 554–63.

9. Hamburger and Hamilton (1992) Hamburger V.and Hamilton H.L. (1951) A serie of normal stage in the development of the chick embryo. *Dev. Dyn.* **195**, 231–272

10. Scaal M, Gros J, Lesbros C, Marcelle C. (2004) In ovo electroporation of avian somites. *Dev Dyn.* **229**, 643–50.

11. Gros J, Serralbo O, Marcelle C. (2009) WNT11 acts as a directional cue to organize the elongation of early muscle fibres. *Nature* **457**, 589–93.

12. Rios AC, Denans N, Marcelle C. (2010) Real-time observation of Wnt beta-catenin signaling in the chick embryo. *Dev. Dyn.* **239**, 346–53.

Chapter 10

Activation and Function of Small GTPases Rho, Rac, and Cdc42 During Gastrulation

Courtney Mezzacappa, Yuko Komiya, and Raymond Habas

Abstract

Gastrulation is comprised of a series of cell polarization and directional cell migration events that establish the physical body plan of the embryo. One of the major ligand-based pathways that has emerged to play crucial roles in the regulation of gastrulation is the non-canonical Wnt signaling pathway. This aspect of Wnt signaling is comprised of a number of signaling branches that are subsequently integrated for the regulation of changes to the actin cytoskeleton during cell polarization and cell migration during vertebrate gastrulation. The Rho family of small GTPases are activated and required for non-canonical Wnt signaling during gastrulation, and in this chapter, we describe biochemical assays for the detection of Wnt-mediated activation of Rho, Rac, and Cdc42 in both mammalian cells and *Xenopus* embryo explants.

Key words: Non-canonical Wnt signaling, Rac, Rho, Cdc42, Gastrulation, *Xenopus*

1. Introduction

Ligand-based signaling pathways play central roles during development of the early embryo and intensive efforts are directed towards elucidating their mechanism of function. The Wnt signaling pathway is an evolutionarily conserved pathway that regulates critical aspects of cell fate determination, cell polarity, cell migration, neural patterning and organogenesis during embryonic development (1). This large family of secreted glycoproteins binds to the seven-pass transmembrane receptor Frizzled (Fz) in conjunction with other co-receptors to regulate a plethora of cellular processes (2). Upon the binding of Wnt protein to its receptor complex, the Wnt signal is transduced to the cytoplasmic protein Dishevelled (Dsh/Dvl) that focuses signaling into two basic branches of Wnt signaling, namely the canonical (β-catenin dependent) and non-canonical (β-catenin independent) pathways. Although Wnt signaling

Kursad Turksen (ed.), *Planar Cell Polarity: Methods and Protocols*, Methods in Molecular Biology, vol. 839,
DOI 10.1007/978-1-61779-510-7_10, © Springer Science+Business Media, LLC 2012

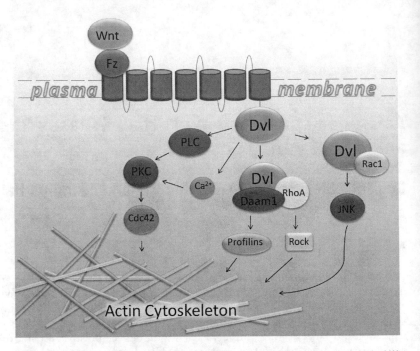

Fig. 1. *A schematic representation of actin cytoskeletal regulation via non-canonical Wnt signaling*. Non-canonical Wnt signaling is defined as Wnt/Frizzled (Fz) initiated signaling independent of β-catenin transcriptional function. The Wnt signal is transduced through the Fz receptor, activating the assembly of different Dvl/effector complexes, ultimately leading to the activation of different pathways regulating the actin cytoskeleton and cell adhesion. A hallmark of non-canonical Wnt signaling is the activation of the small GTPases Rho and Rac. In turn, this activates Rho Kinase (Rock) and Jun N-terminal kinase (JNK), leading to the modulation of the cytoskeleton for cell movement and polarity. For the Wnt/Ca²⁺ pathway, Wnt signaling via Fz mediates the activation of G-proteins to activate PLC, PKC, and Cdc42 for roles in actin dynamics.

via the canonical β-catenin pathway has been most intensively studied in cell fate determination, cell proliferation and human cancer (1, 3), Wnt activation of the Rho family of GTPases via the non-canonical pathway has recently received increased attention (Fig. 1). In addition to the activation of β-catenin and the small GTPases, Wnt proteins can also stimulate less-defined non-canonical signaling pathways such as the Wnt/Ca²⁺, Wnt/Rap, Wnt/ROR, Wnt/PKA, Wnt/aPKC, Wnt/Ryk, and Wnt/mTOR pathways (4).

The Wnt signaling pathway was first identified in *Drosophila melanogaster*, and was later shown to be conserved from nematodes to mammals (5). The canonical pathway was the first branch of Wnt signaling to be identified and remains the most extensively characterized (6). The hallmark of the canonical Wnt pathway is the accumulation and translocation of β-catenin into the nucleus for the regulation of gene transcription (7). The non-canonical pathway functions independently of β-catenin-mediated transcription and regulates actin cytoskeleton organization through the activation of the small GTPases Rho, Rac, and Cdc42. The non-canonical pathway or planar cell polarity (PCP) pathway can be divided into several different branches of which different downstream targets are activated (8) (Fig. 1). It is important to note that crosstalk between the canonical and non-canonical pathways exist as the Rac GTPase was recently shown to modulate canonical signaling pointing to levels of signaling complexity that remain to be unraveled (9).

During early embryogenesis, canonical Wnt signaling plays a pivotal role in cell fate determination, while the non-canonical Wnt pathway is essential for cell polarity and migration during vertebrate gastrulation. During gastrulation, mesodermal cells intercalate along the mediolateral axis, resulting in mediolateral narrowing (convergence) and anteroposterior elongation (extension), ultimately resulting in the highly dynamic process of convergent extension (10, 11). The driving force for convergent extension comes from polarized cell protrusions (lamellipodia), stabilization of these protrusions and migration in the dorsal marginal zone of embryo driven by organized cytoskeletal changes (12, 13). The non-canonical pathway was shown to regulate these convergent extension movements and this pathway shares some common components of the PCP pathway in *Drosophila* (14).

In the non-canonical pathway, Wnts bind to the receptor complex and induce activation of Dvl (15). Activated Dvl is then able to transduce signaling into pathways that lead to the activation of the small GTPases Rho, Rac, and Cdc42 (Fig. 1). The small GTPase Cdc42 may also be activated via the Fz receptor independent of Dvl (16). The Rho family of small GTPases acts as molecular switches that cycle between an active (GTP-bound) and an inactive (GDP-bound) conformation under the control of guanine nucleotide exchange factors (GEFs) and GTPase-activating proteins (GAPs) (6, 17, 18). Rho, Rac, and Cdc42 are responsible for the regulation, assembly, and organization of the actin cytoskeleton in eukaryotic cells. Rho controls the assembly of actin/myosin filaments to generate contractile forces, while Rac and Cdc42 promote actin polymerization at the cell periphery to generate protrusive forces, in the form of lamellipodia and filopodia, respectively (19). The first pathway of activation of the small GTPases in non-canonical Wnt signaling involves Daam1, a protein that binds to the PDZ domain of Dvl. Activation of Rho GTPase leads to the activation of the Rho-associated kinase (ROCK) and myosin, which leads to modification of the actin cytoskeleton and cytoskeletal rearrangement (20–22). During the activation of Rho via Daam1-mediated signaling, the activation of at least one Rho GEF has been identified thus far, WGEF (23), though other GEFs and GAPs remain unknown. The second pathway requires the DEP domain of Dvl and activates Rac GTPase, which in turn stimulates Jun N-terminal Kinase (JNK) activity, independent of Daam1 (20, 24). Wnt/Fz signaling also activates another branch of the non-canonical Wnt pathway called the Wnt/Ca²⁺ pathway. In this case, Wnt/Fz stimulation leads to the release of intracellular Ca^{2+} through trimeric G proteins. Ca^{2+} can then activate protein kinase C (PKC), which regulates cell–cell contact during convergent extension via activation of Cdc42 (6, 25) (Fig. 1).

It is accepted that the Frizzled (Fz) family of serpentine transmembrane receptors are cell surface receptors for Wnt proteins (26). There are 19 Wnt genes and 10 Fz genes in both the mouse and human genomes (27). In addition to Fz, several other families of cell surface receptors, including LDL receptor-related proteins 5 and 6 (LRP5/6) (28), Ryk (29), and proteoglycans (30, 31), are also important for Wnt signal transduction, highlighting the complexity and versatility of the Wnt signaling system. While LRP5/6 were identified as co-receptors for Fz proteins and required specifically for the Wnt/β-catenin pathway (28), a recent study has implicated LRP6 in regulating convergent extension movements during gastrulation (32). Additionally, proteoglycans appear to be involved in extracellular Wnt ligand transport and distribution, thus potentially affect many Wnt pathways (33). Attempts have been made to classify Wnt molecules into either the canonical subfamily (Wnt-1, Wnt-3a, and Wnt-8) or the non-canonical subfamily (Wnt-4, Wnt-5a, and Wnt-11) (6). However, accumulating evidence suggests that such a classification may be oversimplified, and many Wnt proteins, including Wnt-1, Wnt-3a, Wnt-5a, and Wnt-11, can activate multiple pathways in different experimental contexts, likely depending upon receptor complements (including Fz and other receptors) and other cofactors that these Wnt proteins may interact with (20, 22, 34–37). Despite these caveats, it is now well established that activation of the Rho family of GTPases by Wnt/Fz signaling is critical for vertebrate development (10, 11, 38).

Unlike the PCP pathway in *Drosophila* in which the Wnt ligand has not yet been identified, the major ligand regulating convergent extension in *Xenopus* is believed to be Wnt-11 with corresponding receptor as Fz7 (14). It has been reported that Rho and Rac cooperate with Wnt/Fz to promote different important aspects of the process of convergent extension. Wnt-11 activation of RhoA and Rac can be demonstrated in dorsal embryo explants (Figs. 2 and 3), in which interfering with Wnt-11 or Fz function prevents RhoA and Rac activation (20, 39). Rho activation was blocked after expression of dominant-negative Wnt-11, an extracellular fragment of Fz7, or an inhibitor of Dvl. Conversely, overexpression of Wnt-11 or *Xenopus* Fz7 (Xfz7) in the embryo's ventral region, which is not involved in CE movements and does not express Wnt-11 or Xfz7, is sufficient to activate RhoA and Rac (20, 39). Wnt/Fz activation of Rho and Rac has also been observed in several

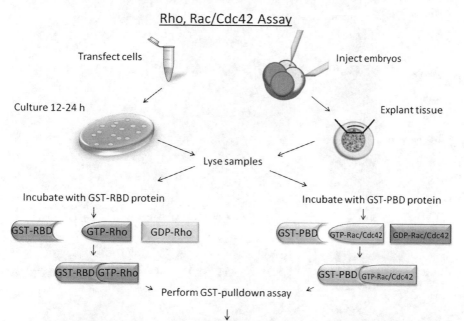

Fig. 2. *A schematic overview of the Rho, Rac, and Cdc42 assays.* For *in vitro* experiments, cells are transfected and cultured for 12–24 h and are subsequently lysed for binding studies. For *in vivo* experiments, *Xenopus* embryos are injected with RNA at Stage 3 and explants are removed at Stage 10.5 to be lysed and incubated with GST-proteins. Samples are then incubated with either GST-RBD or PBD proteins and a GST-pulldown assay is performed. Samples are then subject to SDS-PAGE and Western blotting.

commonly used mammalian cell lines (20, 34, 39). In these experiments (Fig. 2), transfection of Wnt-1 or Wnt-3a cDNA, or of certain (but not all) Fz cDNAs, or treatment with Wnt-1 or Wnt-3a conditioned medium, results in RhoA and Rac activation (20, 34, 39). These observations in vertebrate embryos and mammalian cells followed and paralleled earlier genetic studies in *Drosophila*, which demonstrated that Fz function in PCP relies on RhoA and Rac gene function (40). Thus both genetically and biochemically, Wnt/Fz signaling to RhoA and Rac is a highly evolutionarily conserved function. Likewise, there are some reports that demonstrated Cdc42 plays an important role in convergent extension (6). The expression of either dominant-negative or wild-type Cdc42, in either the animal pole or marginal zone of four-cell embryos, inhibited convergent extension movements (16). Activation of Cdc42 by Wnt-11 is dependent on $G\beta\gamma$ (subunits of heterotrimeric G proteins) acting through protein kinase C (PKC) (41).

In this chapter, we describe biochemical assays used to investigate the activation of Rho, Rac, and Cdc42 GTPases in mammalian culture cells and in the *Xenopus* embryo. These assays utilize a gluthathione *S*-transferase (GST)-pulldown strategy using fusion

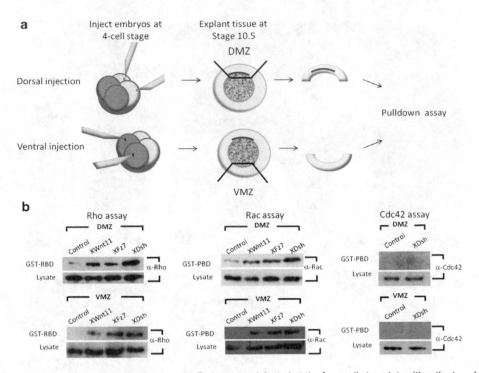

Fig. 3. *Rho, Rac, and Cdc42 assays in Xenopus*. (**a**) Embryos are injected at the four-cell stage into either the two dorsal or two ventral cells. Embryos are then allowed to develop until stage 10.5 for explants. At this point, either the dorsal marginal zone (DMZ) or ventral marginal zone (VMZ) is cut and subjected to Rho or Rac and Cdc42 pulldown assays. (**b**) Examples of Rho, Rac, and Cdc42 assays performed with the DMZs and VMZs of *Xenopus* embryos. Rho, Rac, and Cdc42 activation detected via GST-RBD or GST-PBD pulldown samples show Rho and Rac, but not Cdc42, are activated preferentially in the DMZ. Endogenous Rho, Rac, and Cdc42 levels are shown in the lysate samples.

proteins that specifically bind to the activated/GTP-bound forms of Rho, Rac, and Cdc42. For Rho assays, a Rho-binding fragment of the Rho-effector Rhotekin is fused to GST and termed GST-RBD (42) (Fig. 2). For the Rac and Cdc42 assays, the Rac and Cdc42 binding fragment of p21 (PAK) is fused to GST and termed GST-PBD (43, 44) (Fig. 2). The GST-RBD and GST-PBD fusion proteins are produced in bacterial cells, purified and incubated with cell lysates derived from either mammalian cells or *Xenopus* embryo explants (Fig. 2). GST-RBD and GST-PBD bind specifically to the GTP-bound forms of Rho, Rac, or Cdc42, respectively, which are precipitated using Glutathione-agarose beads and detected by conventional immunoblotting (Figs. 2 and 3). The following protocols provide an efficient method to study the activation of the small GTPases Rho, Rac, and Cdc42 both *in vitro* and *in vivo*.

2. Materials

2.1. Bacteria

BL21 bacterial cells

1M IPTG

LB-ampicillin plates: 1% Bacto tryptone, 0.5% yeast extract, 1% NaCl, 100 mg/ml ampicillin, 1.5% Bacto Agar

LB-ampicillin growth media: 1% Bacto tryptone, 0.5% yeast extract, 1% NaCl, 100 microgram/ml ampicillin

1× PBS: 1.54 mM KH_2PO_4, 155.17 mM NaCl, 2.71 mM $Na_2HPO_4 \cdot 7H_2O$ (pH 7.2)

Glutathione sepharose beads

1× PBS/10 mM DTT/1% TritonX 100

Protease inhibitor cocktail

50 mg/ml lysozyme; 10% Triton-X 100

1 M $MgCl_2$

10 mg/ml DNase1

2.2. Mammalian Cells

HEK293T cells

10% fetal bovine serum (FBS) in Dulbecco's Modified Eagle Medium (DMEM) supplemented with penicillin and streptomycin

Ca^{2+} phosphate

1× Trypsin

1× PBS

2.3. Embryos

Xenopus laevis embryos

10× MMR: 1 M NaCl, 20 mM KCl, 20 mM $CaCl_2$, 10 mM $MgCl_2$, 50 mM Hepes, pH to 7.6

3% Ficoll-0.5× MMR

1% BSA/0.5× MMR

2.4. GST-PRD and GST-PBD Binding Assay Buffers

Rho Lysis Buffer: 50 mM Tris–HCl pH 7.2, 500 mM NaCl, 1% Triton-X 100, 0.5% Sodium deoxycholic acid, 0.1% SDS, 10 mM $MgCl_2$ and 1× protease inhibitors (added fresh each time).

Rho Wash Buffer: 50 mM Tris–HCl pH 7.2, 1% Triton-X 100, 150 mM NaCl, 10 mM $MgCl_2$ and 1× protease inhibitors.

Rac and Cdc42 Lysis Buffer: 50 mM Tris, pH 7.5, 200 mM NaCl, 2% NP40, 10% glycerol, 10 mM $MgCl_2$ and 1× protease inhibitors.

Rac and Cdc42 Wash Buffer: 25 mM Tris, pH 7.5, 40 mM NaCl, 1% NP40, 30 mM $MgCl_2$ and 1× protease inhibitors.

2.5. Western Blotting	12% SDS-PAGE gel

Running Buffer: 25 mM Tris, 192 mM Glycine, 0.1% SDS

Transfer Buffer: 25 mM Tris, 192 mM Glycine, 20% Methanol

1× PBST: 1× PBS, 0.5% Tween-20

2× sample buffer; 125 mM Tris–HCl (pH 6.8), 10% 2-mercapto-ethanol, 4% SDS, 20% Glycerol

5% Non-fat dry milk

Rho monoclonal and polyclonal antibodies (Santa Cruz)

Rac/Cdc42 monoclonal antibodies (Transduction Labs)

SuperSignal PicoWest ECL (Pierce)

3. Methods

3.1. Rho, Rac, and Cdc42 Assays

3.1.1. Preparation of Recombinant GST-RBD Protein

1. Grow an overnight culture of a single colony of BL21 bacterial cells containing the GST-PBD plasmid in 20 ml of LB-amp (100 µg/ml) at 30°C.

2. Dilute the culture into 1 l of LB-amp (100 µg/ml) and grow at 30°C until the optical density at 600 nm is 1.0. This takes 5–7 h depending on the starting optical density.

3. Induce the bacterial culture with 1 ml of 1 M IPTG and incubate for 3–4 h at 30°C.

4. Aliquot the bacteria into 50 ml Falcon tubes and spin at $1,500 \times g$ for 10 min to pellet the bacteria (see Notes 1 and 2). Discard supernatant and flash freeze the pellets in liquid nitrogen.

5. Store the pellets at –80°C. The pellets are stable for up to 1 year.

3.1.2. Preparation of Recombinant GST-PBD Fusion Protein

1. Grow an overnight culture of a single colony of BL21 bacterial cell containing the GST-PBD plasmid in 20 ml LB-amp (100 µg/ml) at 30°C.

2. Dilute the culture into 1 l of LB-amp (100 µg/ml) and grow at 30°C until the optical density at 600 nm is 1.0. This takes 5–7 h depending on starting optical density.

3. Induce the bacterial culture with 1 ml of 1M IPTG and incubate for 3–4 h at 30°C.

4. Lyse the cells in 1× PBS with protease inhibitors using either sonication or the French press method.

5. Pellet the lysate by spinning at $16,000 \times g$ for 15 min and isolate the supernatant.

6. Aliquot the supernatant into 1.5 ml eppendorf tubes and flash freeze in liquid nitrogen (see Notes 1 and 2).

7. Store at –80°C.

3.1.3. Extraction of GST-RBD and GST-PBD

GST-RBD

1. Prepare the Glutathione Sepharose Beads by swelling approximately 100 µl with 1× PBS/10 mM DTT/1% Triton-X 100 for at least 30 min on ice, then wash three times with 500 µl of 1× PBS/10 mM DTT/1% Triton-X 100. After the final wash, the beads can be stored on ice as a 1× slurry. Do not spin beads higher than 800×g during the pelleting and washing stages since this will damage the beads. Also prepare two tubes of beads, for you will have 2 ml of lysate (see Note 3).

2. Thaw one aliquot of frozen GST-RBD pellet on ice and resuspend in 2 ml 1× PBS.

3. Add 20 µl 1 M DTT, 20 µl Protease inhibitor cocktail (Boehringer-Mannheim) and 40 µl of 50 mg/ml lysozyme.

4. Vortex briefly to mix well and incubate on ice for 30 min.

5. Add 225 µl 10% Triton-X 100, 22.5 µl 1 M $MgCl_2$ and 22.5 µl of 10 mg/ml DNase1.

6. Vortex briefly to mix well and incubate on ice for 30 min.

7. Spin at 16,000×g at 4°C for 2 min, and add 1 ml supernatant to each of the two tubes of the pre-swollen beads.

8. Incubate on a Nutator at 4°C for 45 min (do not exceed 1 h).

9. Spin and wash beads three times with 500 µl of 1× PBS/10 mM DTT/1% Triton-X 100.

10. After the final wash, store on ice in 1× slurry with the final volume approximately 500 µl.

GST-PBD

1. Prepare the Glutathione Sepharose Beads by swelling approximately 100 µl with 1× PBS/10 mM DTT/1% Triton-X 100 for at least 30 min on ice. After this, wash three times with 500 µl of 1× PBS/10 mM DTT/1% Triton-X 100 and after the final wash store on ice as a 1× slurry. Do not spin beads higher than 800×g during the pelleting and washing stages in order to not damage the beads. Prepare two tubes of beads for the 1 ml of bacterial lysate (see Note 3).

2. Thaw one aliquot of frozen bacterial supernatant on ice.

3. Add 500 µl supernatant to each of the two tubes of the pre-swollen beads.

4. Incubate on a Nutator at 4°C for 45 min and do not exceed 1 h.

5. Spin and wash beads three times with 500 µl of 1× PBS/10 mM DTT/1% Triton-X 100. After the final wash, store on ice in 1× slurry with the final volume approximately 500 µl.

3.1.4. Sample Preparations for Pulldown Assays

Mammalian Cells

Rho Assay:

1. Mammalian HEK293T cells are cultured in 6-well plates (30 mm) in 10% fetal bovine serum and DMEM media supplemented penicillin/streptomycin until 60% confluency.

2. 1 µg of cDNA is transfected into the cells using standard Ca^{2+}-Phosphate method.

3. Media is changed 12 h post-transfection and cells are incubated for 12–24 h.

Rac and Cdc42 Assay:

1. Mammalian HEK293T cells are cultured in 6-well plates (60 mm) in 10% fetal bovine serum and DMEM media supplemented penicillin/streptomycin until 60% confluency. At this point, the cells are changed into 0.5% fetal bovine serum and DMEM media supplemented penicillin/streptomycin. This step helps to reduce the basal level of activated Rac and Cdc42.

2. Six hours after media is changed to 0.5% sera, 1 µg of cDNA is transfected into the cells using standard Ca^{2+}-Phosphate method.

3. Media is changed 12 h post-transfection and cells are incubated for an additional 12–24 h in 0.5% fetal bovine serum and DMEM media supplemented penicillin/streptomycin.

Xenopus Embryos and Explants:

1. *Xenopus* embryos are injected at the four-cell stage into the two dorsal cells (for dorsal marginal zone [DMZ] explants) or into the two ventral cells (for ventral marginal zone [VMZ] explants) in 3%Ficoll/0.5× MMR.

2. Two hours after injections, embryos are changed into 0.1× MMR and cultured to Stage 10.5.

3. The vitelline membranes are removed from the embryo and the DMZ or VMZ is explanted using forceps. Explants are pooled and stored on ice until they are lysed. All embryos are dissected on agarose-coated culture dishes in a solution of 0.5× MMR/1%BSA. *Xenopus* embryos and explants are handled as described elsewhere (45).

3.1.5. GST-RBD and GST-PBD Binding Assay

1. Lyse the cells in 500 µl of Rho or Rac lysis buffer and to each sample add 10 µl of 10 mg/ml DNase1 solution. Incubate on ice for 10 min and then spin samples at 16,000 ×*g* at 4°C. For *Xenopus* explants, we typically use 50 explants (DMZ or VMZ) for each sample.

2. Remove 25 µl of supernatant and add 25 µl of 2× sample buffer, heat at 90°C for 5 min and store. This is your whole cell lysate for control Western blotting.

3. Remove the remaining 475 µl of supernatant and add to tubes containing approximately 50 µl of GST-beads coupled to the RBD or to PBD.

4. Incubate on a Nutator at 4°C for 1 h and wash three times with Rho or Rac wash buffer, accordingly. After final wash,

resuspend in 50 µl of 2× sample buffer and heat at 90°C for 5 min.

5. Perform Western blot analysis.

3.2. Western Blot Analysis

1. Resolve samples on a 12% SDS-PAGE gel and run until the Bromophenol dye is approximately 1 inch from the bottom of the gel.

2. Transfer to nitrocellulose and block for 1 h with 5% non-fat dry milk for 1 h at room temperature.

3. Wash twice for 5 min with 1× PBST.

4. Incubate with the primary antibody (Rho monoclonal, Santa Cruz for mammalian cell extracts or Rho polyclonal, Santa Cruz, for *Xenopus* explant extracts, Rac or Cdc42 monoclonal, Transduction Labs for both mammalian cell and *Xenopus* explant extracts) at a 1/500 dilution for 1 h at room temperature or overnight at 4°C.

5. Wash once for 15 min and four times for 5 min with 1× PBST.

6. Incubate with secondary antibody at 1/5,000 solution for 1 h at room temperature.

7. Wash once for 15 min and four times for 5 min with 1× PBST.

8. Perform ECL reaction. We typically use SuperSignal PicoWest (Pierce).

9. The endogenous Rho, Rac, or Cdc42 is detectable within 1 min in total lysates or 5 min in the GST-RBD/PBD pull-down samples for mammalian cell extracts. For *Xenopus* explant extracts, the endogenous Rho, Rac, and Cdc42 are detectable within 3–5 min in lysates or 10 min in the GST-RBD/PBD pulldown samples.

4. Notes

1. For each preparation of GST-RBD or GST-PBD bacteria, use one aliquot and follow the protocol to extract the fusion protein and resolve a fraction of the sample on a 12% SDS-PAGE gel and Coomassie Blue stain. GST-RBD runs at approximately 35 kD and GST-PBD approximately 37 kD. If there is extensive degradation, redo the protein preparation.

2. Each 50-ml bacterial aliquot of GST-RBD yield approximately 200–300 µg of fusion protein and each is enough for ten samples for the GST-RBD assay. Each 1-ml aliquot of GST-PBD yields approximately 150–250 µg of protein and each is enough for ten samples for the GST-PBD assay.

3. Always keep samples on ice whenever possible and do not exceed any of the incubation times.

References

1. Logan, C.Y. and R. Nusse, The Wnt signaling pathway in development and disease. Annu Rev Cell Dev Biol, 2004. **20**: p. 781–810.

2. Komiya, Y. and R. Habas, Wnt signal transduction pathways. *Organogenesis*, 2008. **4**(2): p. 68–75.

3. Giles, R.H., J.H. van Es, and H. Clevers, Caught up in a Wnt storm: Wnt signaling in cancer. *Biochim Biophys Acta*, 2003. **1653**(1): p. 1–24.

4. Semenov, M.V., et al., SnapShot: Noncanonical Wnt Signaling Pathways. *Cell*, 2007. **131**(7): p. 1378.

5. Funato, Y., et al., Nucleoredoxin regulates the Wnt/planar cell polarity pathway in *Xenopus*. *Genes Cells*, 2008. **13**(9): p. 965–75.

6. Schlessinger, K., A. Hall, and N. Tolwinski, Wnt signaling pathways meet Rho GTPases. *Genes Dev*, 2009. **23**(3): p. 265–77.

7. Fuerer, C., R. Nusse, and D. Ten Berge, Wnt signalling in development and disease. Max Delbruck Center for Molecular Medicine meeting on Wnt signaling in Development and Disease. *EMBO Rep*, 2008. **9**(2): p. 134–8.

8. Simons, M. and M. Mlodzik, Planar cell polarity signaling: from fly development to human disease. *Annu Rev Genet*, 2008. **42**: p. 517–40.

9. Wu, X., et al., Rac1 activation controls nuclear localization of beta-catenin during canonical Wnt signaling. *Cell*, 2008. **133**(2): p. 340–53.

10. Keller, R., Shaping the vertebrate body plan by polarized embryonic cell movements. *Science*, 2002. **298**(5600): p. 1950–4.

11. Wallingford, J.B., S.E. Fraser, and R.M. Harland, Convergent extension: the molecular control of polarized cell movement during embryonic development. *Dev Cell*, 2002. **2**(6): p. 695–706.

12. Roszko, I., A. Sawada, and L. Solnica-Krezel, Regulation of convergence and extension movements during vertebrate gastrulation by the Wnt/PCP pathway. *Semin Cell Dev Biol*, 2009. **20**(8): p. 986–97.

13. Wallingford, J.B., Planar cell polarity, ciliogenesis and neural tube defects. *Hum Mol Genet*, 2006. **15 Spec No 2**: p. R227–34.

14. Modarresi, R., et al., WNT/beta-catenin signaling is involved in regulation of osteoclast differentiation by human immunodeficiency virus protease inhibitor ritonavir: relationship to human immunodeficiency virus-linked bone mineral loss. *Am J Pathol*, 2009. **174**(1): p. 123–35.

15. Wallingford, J.B. and R. Habas, The developmental biology of Dishevelled: an enigmatic protein governing cell fate and cell polarity. *Development*, 2005. **132**(20): p. 4421–36.

16. Choi, S.C. and J.K. Han, Xenopus Cdc42 regulates convergent extension movements during gastrulation through Wnt/Ca2+ signaling pathway. *Dev Biol*, 2002. **244**(2): p. 342–57.

17. Raftopoulou, M., et al., Regulation of cell migration by the C2 domain of the tumor suppressor PTEN. *Science*, 2004. **303**(5661): p. 1179–81.

18. Raftopoulou, M. and A. Hall, Cell migration: Rho GTPases lead the way. *Dev Biol*, 2004. **265**(1): p. 23–32.

19. Jaffe, A.B. and A. Hall, Rho GTPases: biochemistry and biology. *Annu Rev Cell Dev Biol*, 2005. **21**: p. 247–69.

20. Habas, R., I.B. Dawid, and X. He, Coactivation of Rac and Rho by Wnt/Frizzled signaling is required for vertebrate gastrulation. *Genes Dev*, 2003. **17**(2): p. 295–309.

21. Jessen, J.R., et al., Zebrafish trilobite identifies new roles for Strabismus in gastrulation and neuronal movements. *Nat Cell Biol*, 2002. **4**(8): p. 610–5.

22. Kishida, S., H. Yamamoto, and A. Kikuchi, Wnt-3a and Dvl induce neurite retraction by activating Rho-associated kinase. *Mol Cell Biol*, 2004. **24**(10): p. 4487–501.

23. Tanegashima, K., H. Zhao, and I.B. Dawid, WGEF activates Rho in the Wnt-PCP pathway and controls convergent extension in Xenopus gastrulation. *Embo J*, 2008. **27**(4): p. 606–17.

24. Simons, M., et al., Electrochemical cues regulate assembly of the Frizzled/Dishevelled complex at the plasma membrane during planar epithelial polarization. *Nat Cell Biol*, 2009. **11**(3): p. 286–94.

25. Kohn, A.D. and R.T. Moon, Wnt and calcium signaling: beta-catenin-independent pathways. *Cell Calcium*, 2005. **38**(3–4): p. 439–46.

26. Huang, H.C. and P.S. Klein, The Frizzled family: receptors for multiple signal transduction pathways. *Genome Biol*, 2004. **5**(7): p. 234.

27. Willert, J., et al., A transcriptional response to Wnt protein in human embryonic carcinoma cells. *BMC Dev Biol*, 2002. **2**: p. 8.

28. He, X., et al., LDL receptor-related proteins 5 and 6 in Wnt/beta-catenin signaling: arrows point the way. *Development*, 2004. **131**(8): p. 1663–77.

29. Lu, W., et al., Mammalian Ryk is a Wnt coreceptor required for stimulation of neurite outgrowth. *Cell*, 2004. **119**(1): p. 97–108.

30. Ohkawara, B., et al., Role of glypican 4 in the regulation of convergent extension movements during gastrulation in *Xenopus laevis*. *Development*, 2003. **130**(10): p. 2129–38.

31. Topczewski, J., et al., The zebrafish glypican knypek controls cell polarity during gastrulation movements of convergent extension. *Dev Cell*, 2001. **1**(2): p. 251–64.

32. Tahinci, E., et al., Lrp6 is required for convergent extension during *Xenopus* gastrulation. *Development*, 2007. **134**(22): p. 4095–106.

33. Lin, X., Functions of heparan sulfate proteoglycans in cell signaling during development. *Development*, 2004. **131**(24): p. 6009–21.

34. Endo, Y., et al., Wnt-3a-dependent cell motility involves RhoA activation and is specifically regulated by dishevelled-2. *J Biol Chem*, 2005. **280**(1): p. 777–86.

35. He, X., et al., A member of the Frizzled protein family mediating axis induction by Wnt-5A. *Science*, 1997. **275**(5306): p. 1652–4.

36. Qiang, Y.W., et al., Wnt signaling in B-cell neoplasia. *Oncogene*, 2003. **22**(10): p. 1536–45.

37. Tao, Q., et al., A novel G protein-coupled receptor, related to GPR4, is required for assembly of the cortical actin skeleton in early *Xenopus* embryos. *Development*, 2005. **132**(12): p. 2825–36.

38. Veeman, M.T., et al., Zebrafish prickle, a modulator of noncanonical Wnt/Fz signaling, regulates gastrulation movements. *Curr Biol*, 2003. **13**(8): p. 680–5.

39. Habas, R., Y. Kato, and X. He, Wnt/Frizzled activation of Rho regulates vertebrate gastrulation and requires a novel Formin homology protein Daam1. *Cell*, 2001. **107**(7): p. 843–54.

40. Mlodzik, M., Planar cell polarization: do the same mechanisms regulate Drosophila tissue polarity and vertebrate gastrulation? *Trends Genet*, 2002. **18**(11): p. 564–71.

41. Penzo-Mendez, A., et al., Activation of Gbetagamma signaling downstream of Wnt-11/Xfz7 regulates Cdc42 activity during *Xenopus* gastrulation. *Dev Biol*, 2003. **257**(2): p. 302–14.

42. Ren, X.D., W.B. Kiosses, and M.A. Schwartz, Regulation of the small GTP-binding protein Rho by cell adhesion and the cytoskeleton. *Embo J*, 1999. **18**(3): p. 578–85.

43. Akasaki, T., H. Koga, and H. Sumimoto, Phosphoinositide 3-kinase-dependent and -independent activation of the small GTPase Rac2 in human neutrophils. *J Biol Chem*, 1999. **274**(25): p. 18055–9.

44. Benard, V., B.P. Bohl, and G.M. Bokoch, Characterization of rac and cdc42 activation in chemoattractant-stimulated human neutrophils using a novel assay for active GTPases. *J Biol Chem*, 1999. **274**(19): p. 13198–204.

45. Sive, H.L., Grainger, R.M. and Harland, R.M., Early Development of Xenopus laevis. A laboratory manual. 2000, Cold Spring Harbor, NY: Cold Spring Harbor Press.

Chapter 11

Convergent Extension Analysis in Mouse Whole Embryo Culture

Sophie E. Pryor, Valentina Massa, Dawn Savery, Nicholas D.E. Greene, and Andrew J. Copp

Abstract

Mutations have been identified in a non-canonical Wnt signalling cascade (the planar cell polarity pathway) in several mouse genetic models of severe neural tube defects. In each of these models, neurulation fails to be initiated at the 3–4 somite stage, leading to an almost entirely open neural tube (termed craniorachischisis). Studies in whole embryo culture have identified a defect in the morphogenetic process of convergent extension during gastrulation, preceding the onset of neural tube closure. The principal defect is a failure of midline extension, both in the neural plate and axial mesoderm. This leads to an abnormally wide neural plate in which the elevating neural folds are too far apart to achieve closure. In this chapter, we provide details of several experimental methods that can be used to evaluate convergent extension in cultured mouse embryos. We describe analytical methods that can reveal the abnormalities that characterise neurulation-stage embryos with defective planar cell polarity signalling, in particular the loop-tail (*Lp*; *Vangl2*) mutant.

Key words: Mouse, Neurulation, Whole embryo culture, DiI labelling, Electroporation, Convergent extension, Neural tube defects, Craniorachischisis, Vangl2, Loop-tail

1. Introduction

In the mouse embryo, neural tube closure initiates at the hindbrain-cervical boundary (Closure 1) at embryonic day (E) 8.5 (day 22–23 of gestation in humans) and then progresses bidirectionally into the brain and down the spine. Subsequent closure initiation events occur at the forebrain–midbrain boundary (Closure 2) and at the rostral extremity of the forebrain (Closure 3). Closure progresses simultaneously along the spinal region, culminating in closure of the posterior neuropore (PNP) at E10.5 (26–30 days in humans), marking the completion of primary neurulation. For a review of mammalian neurulation, see ref. 1.

Kursad Turksen (ed.), *Planar Cell Polarity: Methods and Protocols*, Methods in Molecular Biology, vol. 839,
DOI 10.1007/978-1-61779-510-7_11, © Springer Science+Business Media, LLC 2012

Failure of any of these closure events results in an open neural tube defect (NTD). The most severe form of NTD, craniorachischisis, arises when the embryo fails to achieve Closure 1 and combines anencephaly of the midbrain and hindbrain with an entirely open spinal region. While more than 200 genes have been described as necessary for successful neural tube closure in mice (2), the perturbation of a minority of these genes leads to craniorachischisis. To date, all of the well-established models of craniorachischisis have been found to lack functional components of the planar cell polarity (PCP) pathway, establishing a striking connection between the initial event of primary neurulation and non-canonical Wnt signalling.

Closure 1 fails in a number of mutant mice including *loop-tail* (*Lp; Vangl2*) (3, 4), *circletail* (*Crc; Scrb1*) (5), *crash* (*Crsh; Celsr1*) (6) and *protein tyrosine kinase 7* (*Ptk7*) (7), as well as in mice doubly mutant for *dishevelled 1* (*Dvl1*) and *Dvl2* (8), for *Dvl2* and *Dvl3* (9), and for *frizzled 3* (*Frz3*) and *Frz6* (10). In each of these single or double mutants, the neural plate lacks the normal, well-defined midline bending point and instead displays an abnormally broad floor plate precursor. Although the neural folds form and elevate towards the midline, the increased distance between them precludes apposition and fusion. In the *Lp* (*Vangl2*) mouse, it has been shown that this genetic requirement for PCP signalling is mediated through convergent extension (CE) cell movements in the midline neural plate and underlying axial mesoderm, prior to Closure 1. *Vangl2* mutants display a cell-autonomous CE defect, whereby the midline fails to narrow and extend, resulting in the widely spaced neural folds (11, 12).

The *Lp* (*Vangl2*) mutation also increases susceptibility to other NTDs, in addition to craniorachischisis, particularly when in double mutant combinations (13). For example, *Lp* interacts genetically with mutations in both *Ptk7* (7) and *grainyhead-like-3* (*Grhl3*) (14) to produce spina bifida, and with *collagen triple helix repeat containing 1 (Cthrc1)* to produce exencephaly (15). It is not yet understood why a combination of *Vangl2* with other mutants can produce such variable phenotypes. In particular, it is unclear whether the genetic interactions represent defects of convergence and extension. If so, this would implicate CE cell movements as an essential feature of later neurulation events, as well as their recognised key role in gastrulation, pre-Closure 1.

The aim of this chapter is to describe some techniques that can be used with early mouse embryos to assess convergence and extension, and initiation of neural tube closure, in PCP mutants. We describe the process of dissecting and culturing E7.5 pre-neurulation stage embryos, combined with the labelling of the midline either by focal injection of DiI into the node or by electroporation of a green fluorescent protein (GFP) expressing vector into the neural plate. We discuss how to evaluate the distinguishing features of

Closure 1 failure by examining defective CE and axial extension in
Lp mutants, and by morphological examination of the floor plate
and notochord.

2. Materials

2.1. Embryo Dissection

1. Dulbecco's Modified Eagle's Medium containing 25 mM HEPES
   (DMEM, stored at 4°C) (Gibco, UK) and supplemented with
   10% foetal calf serum (FCS). This should be made fresh as
   required and warmed to 37°C for use during embryo collection
   and dissection. Inclusion of HEPES ensures maintenance of
   pH when used on the open bench.

2. Watchmakers' forceps (number 5, Dumont, Switzerland). These
   should be precisely sharpened using fine emery paper to allow
   accurate dissection without applying mechanical stress to the
   embryo. Dip forceps tips in alcohol and briefly flame-sterilise
   using a spirit burner before use.

3. Dissecting scissors. Flame sterilise before use.

4. Zeiss SV6 or similar stereomicroscope with transmitted light
   stage.

**2.2. DiI Injection
into the Node**

1. DiI   (1,1'-dioctadecyl-3,3,3',3'-tetramethylindocarbocyanine
   perchlorate, Cell Tracker™ CM-DiI, Molecular Probes, USA):
   1 mg dissolved in 25 μl dimethyl sulphoxide, then diluted
   tenfold in 3 mol/l sucrose. Store aliquots at –20°C, protected
   from light. Thaw aliquot before use and the DiI solution can then
   be stored at 4°C, protected from light, for several days.

2. Petri dish, 55 mm diameter, containing 4% agarose (in PBS),
   2–3 mm deep.

3. Mouth pipetting tube.

4. Glass micropipette needles, pulled on a Flaming-Brown horizontal pipette puller (Sutter Instrument Co., USA).

**2.3. Electroporation
of GFP-Expressing
Vector into Neural
Plate**

1. DNA solution containing GFP-expression vector. For example,
   pCAβ-*m*GFP6 in which GFP expression is driven by the chick
   β-actin promoter under influence of the cytomegalovirus
   enhancer (16) has proven effective in our hands (11). Prepare
   a solution for electroporation to contain 2.5 mg/ml DNA,
   0.1% Fast Green in PBS-DEPC.

2. Petri dish containing agarose, glass micropipette needles and
   mouth pipetting tube, as above.

3. A pair of gold 5 mm point electrodes (BTX model 508, Harvard
   Apparatus) attached to a BTX ECM830 electroporator.

2.4. Whole Embryo Culture

1. Rat serum. This should be stored in small aliquots (1–5 ml) at –20°C and thawed immediately before use (see Note 1).

2. Culture tubes (30 ml, plastic; Nunc) and Millipore filter (0.45 μm pore size).

3. Glisseal silicone grease (Borer Chemie, Switzerland).

4. Gas mixture (see Note 2).

5. Roller culture incubator (37°C; B.T.C Engineering, Cambridge).

2.5. Assessment of Embryo Phenotype and Gene Expression Analysis After Culture

1. Epifluorescence stereomicroscope (e.g., Leica MZ FLIII) with digital camera.

2. Phosphate buffered saline treated with diethyl pyrocarbonate (PBS-DEPC) and autoclaved.

3. Glass microscope slides for flat mounting.

4. Eyepiece graticule for making linear measurements.

5. Ice-cold paraformaldehyde (PFA, 4%; dissolved in PBS-DEPC) for fixation. This should be stored as aliquots at –20°C and thawed immediately before use with warming to ensure all PFA is dissolved.

6. Standard reagents and materials for in situ hybridisation on whole mount embryos or on histological sections.

2.6. Examination of Morphology and Gene Expression Patterns in Embryo Sections

1. Small glass embedding moulds (Agar Scientific).

2. Histoclear (National Diagnostics).

3. Paraffin wax (Thermo Scientific). This should be melted at 56°C prior to sectioning.

4. Two old pairs of forceps for heating and orientation of specimens.

5. Paper strips for pencil labelling of wax blocks.

6. Agarose (2% in PBS). Dissolve in microwave or water bath and keep warm.

7. Small plastic embedding moulds (Thermo Scientific)

8. Razor blade or scalpel for trimming wax and agarose blocks after embedding.

9. Hematoxylin and eosin for counterstaining.

3. Methods

The ability to culture mouse embryos prior to and during neurulation allows a number of experimental studies and chemical manipulations which would not otherwise be possible in a mammalian system. For successful cultures, carefully dissected embryos and high quality rat serum are essential. Here we describe two methods for vitally labelling mouse embryos in order to study CE: DiI labelling by microinjection into the node tissue, and electroporation of a GFP-expression vector into the neural plate.

3.1. Embryo Dissection

1. Dissect the uterus from the pregnant female and place into a 55 mm petri dish containing pre-warmed dissecting medium (DMEM + 10% FCS). Place the dish on the stereomicroscope stage, using transmitted (under-stage) illumination. It is generally best to use the "dark field" adjustment for the early stages of dissection, and change to "bright field" for later stages when the embryo has been removed.

2. Trim away fat and blood vessels from the mesometrial surface of the uterus (Fig. 1a).

3. Beginning at the mesometrial surface, gently open the uterus at the site of each implantation in turn by creating a hole in the uterine wall. Expand this hole carefully until the decidual swelling protrudes from the elastic uterine wall. Steady the uterus with one pair of forceps while using the other to grip across the width of the uterus next to the decidua, and then gently "milk it out" of the hole. Repeat this process until all of the decidual swellings have been removed from the uterus (Fig. 1b). Transfer the decidual swellings to a fresh dish of dissecting medium.

4. Open each decidua starting at the anti-mesometrial (fluffy) end (Fig. 1b), to reveal the trophoblast layer which envelops the embryo (Fig. 1c). Leave the ectoplacental cone intact and, should the mural trophoblast tear open, take care not to damage the underlying yolk sac. Gently release the conceptus from within the swelling by peeling away the decidual debris (Fig. 1d). Transfer all conceptuses to a fresh dish of dissecting medium.

5. Next, remove the mural trophoblast together with the thin, elastic Reichert's membrane, which is almost invisible beneath the trophoblast. Grasp the trophoblast with both pairs of forceps in order to tear open the membrane and remove it from the outside of the yolk sac. If successful, the trophoblast layer will come off cleanly in a sheet (Fig. 1e). However, if it detaches in pieces then this indicates that Reichert's membrane is still intact. As before, it is important not to rupture the yolk sac.

6. Finally, trim Reichert's membrane up to the edge of the ectoplacental cone (see Note 3). Once all the embryos have been dissected to this point (Fig. 1f), they are ready to be injected with DiI, electroporated or put straight into culture (see Note 4).

3.2. Injection of DiI into the Node

1. Transfer the dissected embryos to a 55 mm agarose petri dish containing fresh dissecting medium. Create a small well in the agarose into which the embryo can be orientated, so as to be viewed from its posterior surface (Fig. 2a). Lodge it firmly in place to facilitate injection.

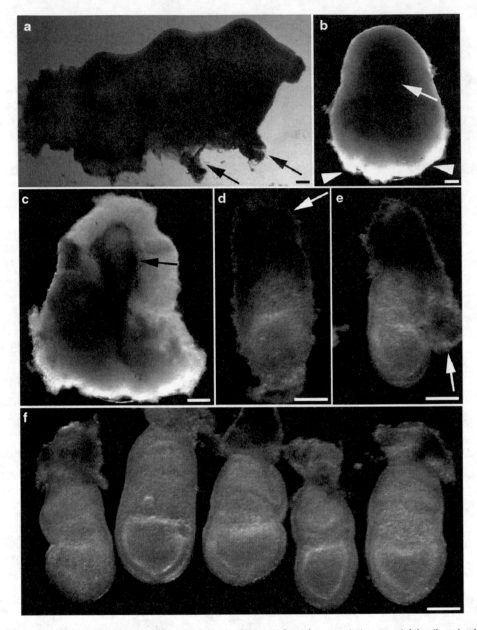

Fig. 1. Dissection of E7.5 mouse embryos for whole embryo culture. (**a**) Part of pregnant uterus containing three implantation sites. *Arrows* indicate fat and blood vessels at the mesometrial surface. (**b**) Decidual swelling after removal from the uterus. The outline of the embryo can be seen as a darker region (*arrow*). The next stage of dissection should begin from the broad, fluffy end (*arrowheads*). (**c**) Decidual swelling dissected open to reveal the conceptus, encased in trophoblast (*arrow*). *Arrowhead* indicates the ectoplacental cone. (**d**) Conceptus after being released from the decidual swelling. Trophoblast cells can be seen on the surface and the ectoplacental cone remains intact (*arrow*). (**e**) Reichert's membrane and the overlying trophoblast (*arrow*) have been partially removed by peeling away from the embryonic region. (**f**) Five dissected embryos after complete removal of Reichert's membrane. The yolk sacs remain intact and the embryos are now ready to be cultured. Scale bars represent: 400 μm in (**a–c**); 200 μm in (**d–f**).

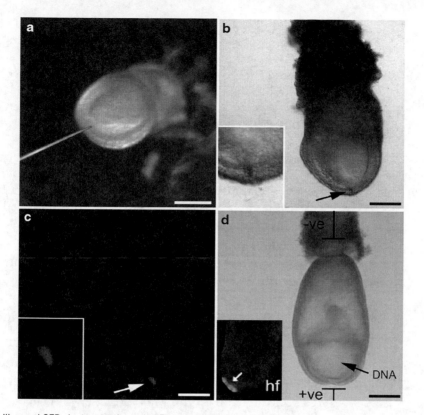

Fig. 2. Dil labelling and GFP electroporation using E7.5 embryos. (**a**) Embryo held in place for Dil injection by positioning the ectoplacental cone in agarose. The microinjection needle indicates the site of injection at the node. (**b**) Bright field image of embryo after Dil injection into the node. A *pink line* of labelled cells can be seen (*arrow* and inset). (**c**) Fluorescence image of embryo in **b**. *Arrow* indicates the Dil-labelled node (enlarged in inset). (**d**) Diagram showing the method for electroporating the neural plate of E7.5 embryos. DNA is injected into the amniotic cavity (*arrow*), after which current is passed between the electrodes, with the anode positioned adjacent to the node region. Inset (reproduced with permission from ref. 11): embryo 4 h after electroporation, showing early GFP expression in the caudal neural plate (*arrow*). Headfold indicated by hf. Scale bars represent 200 μm in (**a–d**).

2. Place a drop of DiI solution onto a petri dish lid or piece of parafilm. Break the injecting needle to create a micropipette. The orifice diameter should be just wide enough to allow DiI solution to be drawn up into the needle, but small enough that the solution does not leak out when the needle is placed into the dissecting medium (see Note 5). Attach the injecting needle to the mouth pipette tube and draw up a small amount of DiI (e.g., 0.5 μl).

3. Insert the needle into the embryonic node, allowing it to just penetrate through into the amniotic cavity. Gently release the DiI while slowly withdrawing the needle.

4. Repeat for the remaining embryos. It is not necessary to accurately control and standardise the volume of DiI that is injected into each embryo; simply ensure that the node is sufficiently

labelled. To check for the correct localisation of DiI prior to culture, embryos can be examined under the light stereomicroscope, in which case a thin line of pink-labelled cells should be seen (Fig. 2b). When viewed using a fluorescence stereomicroscope a strong DiI signal should be seen in the node (Fig. 2c). Do not worry if some DiI solution can also be seen dispersed within the amniotic cavity (as in Fig. 2b).

3.3. Electroporation of GFP-Expression Vector into the Neural Plate

1. Place embryos in PBS-DEPC in a 55 mm agarose petri dish on the stereomicroscope stage and inject the DNA solution containing Fast Green into the amniotic cavity using a hand-held glass micropipette (Fig. 2d). Injection is continued until the amniotic cavity swells slightly.

2. Place the embryo immediately between the gold point electrodes attached to the electroporator, with the ventral midline of the caudal embryonic region next to the anode (Fig. 2d).

3. Pass the current (5 pulses, 50 ms, 15 V), and then repeat for each embryo to be electroporated. Place embryos into culture immediately afterwards.

3.4. Whole Embryo Culture

1. Prepare the rat serum in advance of DiI injection or embryo electroporation. Thaw aliquots by warming to 37°C and then pass through a 0.45 μm Millipore filter. In general we culture two E7.5 embryos per ml of serum, although this will vary depending on the length of culture and embryonic stage.

2. Prepare the culture tubes by smearing the outer rim with a small amount of silicone grease to create an airtight seal.

3. Pipette the required volume of serum into the culture tubes. Gas the serum for 1 min with the appropriate mixture by attaching a pasteur pipette via plastic tubing to the cylinder and gently blowing the gas on to the inner wall of the tube. The surface of the serum should gently ripple without bubbling.

4. Leave the culture tubes at 37°C (at least 15 min) until the embryos are ready. Embryos should be placed into culture as soon as possible after DiI injection or electroporation.

5. Gently transfer the embryos into the serum using a Pasteur pipette, ensuring that a minimal amount of dissecting medium is introduced into the culture tube.

6. Gas the serum again and place the tubes into the roller culture incubator. Cover the lid to protect from light.

7. Culture for 18–20 h (i.e., overnight), rolling. Re-gas every 6–12 h (see Note 6).

3.5. Assessment of Embryonic Phenotype After Culture

E7.5 embryos cultured for 18–20 h will not yet have developed a functioning yolk sac circulation, which is a later indicator of health following a period in culture. Therefore, to evaluate the success of the cultures at this earlier stage, transfer the embryos to fresh, warm dissecting medium and examine their overall appearance. The yolk sac should look intact and round, with a smooth surface, and contain a suitably sized embryo at the early somite stage (Fig. 3a, b). A good strategy is to arrange for a second litter of age-matched embryos to be available for dissection just prior to harvesting embryos from culture. Then compare the "in vivo" embryos with those from culture. Detailed methods for scoring the morphology of embryos at E8.5 are available: (17, 18).

1. Gently open the yolk sac and underlying amnion using forceps and remove from around the embryo (Fig. 3c, d). If the yolk sac is to be kept for genotyping, first remove and discard the remains of the ectoplacental cone, which may have maternal blood contamination. Rinse the yolk sac in PBS-DEPC and place into an eppendorf tube.

2. To flat mount for photography, rinse the embryo in PBS-DEPC and place onto a glass microscope slide. Using forceps, orientate the embryo so that it lies dorsal side up. Carefully soak up some of the PBS from the edge of the droplet with a tissue, until the surface tension "pulls" the embryo into a flat position (Fig. 3e, see Note 7). The developmental stage can now be determined by counting the somites.

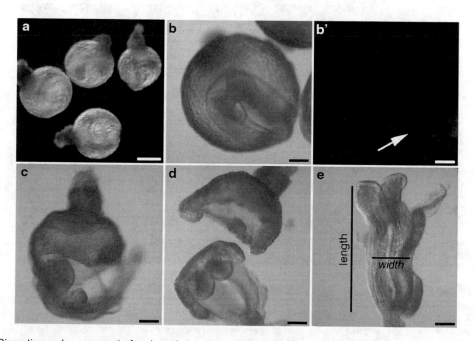

Fig. 3. Dissection and assessment of embryonic length-to-width ratio after culture. (a) Healthy embryos at early somite stage, following 18–20 h in culture. (b, b′) Cultured embryo viewed from the ventral surface (rostral to upper left). *Arrow* in b′ indicates DiI-labelled cells extending along the midline. (c, d) Removal of the yolk sac and ectoplacental cone prior to analysis in cultured embryos. (e) Flat mounted wild-type embryo showing the method for measuring length-to-width ratio. Scale bars represent: 400 μm in A, 200 μm in (b–e).

3. One characteristic of *Lp* (*Vangl2*) mutants is a reduction in the embryonic length-to-width (L:W) ratio due to failure of axial extension. To assess this phenotype, the total length and width of the embryo can be measured after flat mounting using an eyepiece graticule, or later from a photograph (Fig. 3e).

4. To evaluate CE and axial extension, examine the DiI-labelled or GFP-electroporated flat mount embryos under the fluorescence stereomicroscope. In wild-type DiI-labelled embryos, red-fluorescent cells should be observed along the axial midline (Figs. 3b, b' and 4a, a'), representing labelled notochord and floor plate which extend rostral to the node. Similarly, in wild-type GFP-electroporated embryos, green-fluorescent cells should be visible in the midline neural plate from the caudal site of electroporation rostrally into the future brain region (Fig. 4c, d). In contrast, in a large proportion of *Lp* homozygous mutants, labelled cells will persist caudally at the injection or electroporation site due to defective midline extension (Fig. 4b, b' and e). In *Lp* heterozygotes a variable, intermediate phenotype is generally seen.

5. If the embryos are to be processed further, they can be fixed flat on the microscope slide by replacing the PBS with a drop

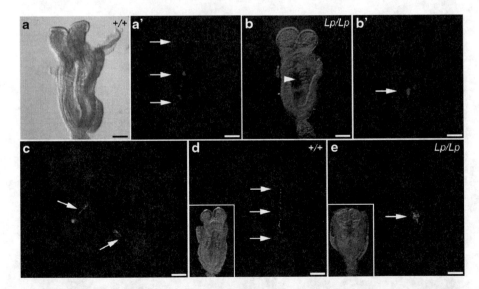

Fig. 4. Assessment of PCP phenotype in DiI-labelled and GFP-electroporated embryos after culture. (**a**, **a'**) Bright field and fluorescence images of flat mounted wild-type embryo. DiI-labelled cells extend rostral to the node (*arrows* in **a'**), indicating normal CE and axial elongation. (**b**, **b'**) Bright field and fluorescence images of a *loop-tail* homozygous mutant. The embryo displays failure of Closure 1 (*arrowhead* in **b** indicates open neural tube) and very limited midline extension of DiI-labelled cells (*arrow* in **b'**). (**c**) Wild-type embryos cultured for 18–20 h following electroporation of a GFP-expression vector into the caudal neural plate. GFP-labelled cells extend along the midline (*arrows*). (**d**, **e**) Fluorescence (inset: bright-field) images of embryos following GFP electroporation. The wild-type embryo (**d**) displays marked extension of GFP-labelled cells along the neural plate midline (*arrows*), whereas the *loop-tail* homozygous embryo (**e**) has undergone minimal axial extension of GFP-positive cells (*arrow*). Scale bars represent: 200 μm in (**a–e**).

of 4% PFA (the fixative and length of fixation depends on the type of further analysis to be carried out).

6. After fixation, rinse the embryos in ice-cold PBS-DEPC, dehydrate through a graded series of methanol washes and store in 100% methanol at –20°C.

3.6. Examination of Morphology and Gene Expression in Neurulation-Stage Embryos

Further evidence of defective CE in *Lp* embryos is provided by the abnormally wide floor plate precursor and notochord (12). To examine this aspect of the phenotype, transverse sections of the cultured embryos can be cut and counterstained with hematoxylin and eosin. An alternative approach is to perform whole mount in situ hybridisation for *sonic hedgehog* (*Shh*) mRNA, a marker of notochord (and floor plate following neural tube closure). The protocol for whole mount in situ hybridisation is as described previously (19), based on the technique of Wilkinson (20). After developing, embryos should be re-fixed and dehydrated to methanol for embedding in paraffin wax (for microtome sectioning), or rinsed in PBS for agarose embedding (for vibratome sectioning). Preparation of transverse histological sections then enables study of floor plate and notochordal morphology.

3.6.1. Paraffin Wax Embedding

1. Remove the embryos from –20°C and wash twice in 100% ethanol for 30 min, rocking at room temperature (RT).

2. Transfer each embryo to a separate glass mould, remove the ethanol and fill the well with Histoclear. Leave at RT for 20 min.

3. Wash again with Histoclear, this time for 20 min in a 60°C oven.

4. Incubate in three changes of molten paraffin wax, 45 min at 60°C each, to allow penetration (see Note 8).

5. At the end of the final wash, view the glass well under a stereomicroscope. Using heated forceps to keep the wax molten, gently move the embryo around until a solid layer of soft wax has formed at the base of the well. Orientate the embryo for transverse sectioning by gently positioning it in the soft layer as desired. Insert a paper label and allow the remaining wax to set overnight at RT.

6. Remove the solid wax block from the mould by placing at –20°C until it can be easily pushed out.

7. Trim the block to produce a cube which can then be attached to a microtome chuck using a small amount of molten wax. Place on ice for 30 min before securing to the microtome for sectioning. Cut 7–12 μm thick sections.

3.6.2. Agarose Embedding

8. Wash embryos in PBS (or rehydrate to PBS if they have been fixed and dehydrated to 100% methanol after in situ hybridisation).

9. Pour some warm agarose (2%; dissolved in PBS) into a plastic embedding mould.

10. Transfer an embryo to the mould (in minimal PBS).

11. As the agarose begins to polymerise, use forceps to move the embryo to the centre of the mould (briefly placing on ice speeds up the polymerisation process). For transverse sectioning, the embryo should be positioned lying on its side.

12. When the embryo remains in position, place the well on ice to allow the agarose to set completely.

13. Remove the agarose block from the well and cut to size. Remember that, when attached to the vibratome, the first sections will be cut from the top of the block, so ensure that the desired region of the embryo is closest to this side.

4. Notes

1. Prepare serum for culture by withdrawing blood from the abdominal aorta of male rats terminally anaesthetised using diethyl ether or isofluorane. The blood is immediately centrifuged (5 min at $1,000 \times g$) to pellet the red cells and then allowed to clot so that the serum can be extracted by gently squeezing the fibrin clot. After heat inactivation, the serum from different rats is pooled to produce a batch of uniform quality. All embryos in an experiment should be exposed to the same serum batch during culture. For a more detailed description of serum preparation, see ref. 21.

2. The required composition of the gas mixture varies depending on the embryonic stage. When culturing E7.5-E9.5 embryos use: 5% O_2, 5% CO_2 and 90% N_2. For E9.5-E10.5 use: 20% O_2, 5% CO_2 and 75% N_2. For embryos older than E10.5 use: 40% O_2, 5% CO_2 and 55% N_2.

3. Removal of Reichert's membrane is the most challenging stage of the dissection process. It may be easiest to grasp the membrane at the opposite side to the ectoplacental cone, stretching it away from the embryo slightly so that it can be torn with the other pair of forceps. Reichert's membrane should be removed as close to the ectoplacental cone as possible as it can become "sticky" during culture, causing the embryo to adhere to the tube surface.

4. Dissection of E7.5 embryos for culture is a delicate process, requiring extreme care. However, it is also important to work as quickly as possible, minimising the length of time between removal of the embryos from the uterus and the start of the culture experiment.

5. The DiI injection needle should be broken so that its end is smooth and perpendicular to the axis of the needle (not bevelled or jagged). This process is most easily achieved using a microforge (e.g., Narishige, Japan), but can be performed adequately by hand, given practice. It is best to prepare a number of injection needles in advance.

6. For overnight cultures we find that the embryos develop well if gassed in the early evening and again first thing the following morning. For shorter cultures (around 6 h) the embryos do not need to be re-gassed after the start of the experiment.

7. When removing the PBS from the slide, take great care not to soak up the embryo as it will become lost on the tissue.

8. The wax will quickly solidify at RT, so try to change the washes as quickly as possible. To maintain the temperature of the moulds, it is helpful to place the embryos on a metal tray inside the oven. This can then be used to transport the moulds to and from the wax container. Pasteur pipettes for changing washes should also be kept at 60°C. If the wax solidifies at any point, simply place the mould back into the oven and continue once it has melted.

Acknowledgements

The authors' research on convergent extension is supported by the Wellcome Trust and Medical Research Council.

References

1. Copp AJ, Greene NDE, Murdoch JN (2003) The genetic basis of mammalian neurulation. *Nat Rev Genet* 4:784–93

2. Harris MJ, Juriloff DM (2007) Mouse mutants with neural tube closure defects and their role in understanding human neural tube defects. *Birth Defects Res A Clin Mol Teratol* 79: 187–210

3. Kibar Z, Vogan KJ, Groulx N et al (2001) *Ltap*, a mammalian homolog of *Drosophila Strabismus/Van Gogh*, is altered in the mouse neural tube mutant Loop-tail. *Nature Genet* 28:251–5

4. Murdoch JN, Doudney K, Paternotte C et al (2001) Severe neural tube defects in the *loop-tail* mouse result from mutation of *Lpp1*, a novel gene involved in floor plate specification. *Hum Mol Genet* 10:2593–601

5. Murdoch JN, Henderson DJ, Doudney K et al (2003) Disruption of *scribble* (*Scrb1*) causes severe neural tube defects in the *circletail* mouse. *Hum Mol Genet* 12:87–98

6. Curtin JA, Quint E, Tsipouri V et al (2003) Mutation of *Celsr1* disrupts planar polarity of inner ear hair cells and causes severe neural tube defects in the mouse. *Curr Biol* 13:1–20

7. Lu X, Borchers AG, Jolicoeur C et al (2004) PTK7/CCK-4 is a novel regulator of planar cell polarity in vertebrates. *Nature* 430:93–8

8. Wang J, Hamblet NS, Mark S et al (2006) Dishevelled genes mediate a conserved mammalian PCP pathway to regulate convergent extension during neurulation. *Development* 133:1767–78

9. Etheridge SL, Ray S, Li S et al (2008) Murine dishevelled 3 functions in redundant pathways

with dishevelled 1 and 2 in normal cardiac outflow tract, cochlea, and neural tube development. *PLoS Genet* 4:e1000259

10. Wang Y, Guo N, Nathans J (2006) The role of Frizzled3 and Frizzled6 in neural tube closure and in the planar polarity of inner-ear sensory hair cells. *J Neurosci* 26:2147–56

11. Ybot-Gonzalez P, Savery D, Gerrelli D et al (2007) Convergent extension, planar-cell-polarity signalling and initiation of mouse neural tube closure. *Development* 134:789–99

12. Greene NDE, Gerrelli D, Van Straaten HWM et al (1998) Abnormalities of floor plate, notochord and somite differentiation in the *loop-tail* (*Lp*) mouse: a model of severe neural tube defects. *Mech Dev* 73:59–72

13. Greene NDE, Stanier P, Copp AJ (2009) Genetics of human neural tube defects. *Hum Mol Genet* 18:R113–R129

14. Stiefel D, Shibata T, Meuli M et al (2003) Tethering of the spinal cord in mouse fetuses and neonates with spina bifida. *J Neurosurg* (Spine) 99:206–13

15. Yamamoto S, Nishimura O, Misaki K et al (2008) Cthrc1 selectively activates the planar cell polarity pathway of Wnt signaling by stabilizing the Wnt-receptor complex. *Dev Cell* 15:23–36

16. Yaneza M, Gilthorpe JD, Lumsden A et al (2002) No evidence for ventrally migrating neural tube cells from the mid- and hindbrain. *Dev Dyn* 223:163–7

17. Brown NA (1990) Routine assessment of morphology and growth: scoring systems and measurements of size. In: Copp AJ, Cockroft DL (eds) Postimplantation Mammalian Embryos: A Prac-tical Approach, pp 93–108, IRL Press, Oxford

18. Van Maele-Fabry G, Delhaise F, Picard JJ (1990) Morphogenesis and quantification of the development of post-implantation mouse embryos. *Toxic in Vitro* 4:149–56

19. Ybot-Gonzalez P, Copp AJ, Greene NDE (2005) Expression pattern of glypican-4 suggests multiple roles during mouse development. *Dev Dyn* 233:1013–7

20. Wilkinson DG (1992) In Situ Hybridisation: A Practical Approach. IRL Press, Oxford

21. Copp A, Cogram P, Fleming A et al (2000) Neurulation and neural tube closure defects. *Methods Mol Biol* 136:135–60

Chapter 12

Analysis of PCP Defects in Mammalian Eye Lens

Yuki Sugiyama and John W. McAvoy

Abstract

Multicellular tissues and organs often show planar cell polarity (PCP) where the constituent cells align along an axis to form coordinated patterns. Mammalian eye lenses are mainly comprised of epithelial-derived fibre cells, which exhibit highly ordered alignment that is regulated by PCP signaling. Each fibre cell has an apically situated primary cilium and in most cases this is polarized towards the lens anterior pole. Here we describe how to visualize the global cellular alignment of lens fibre cells by examining the suture pattern that is formed by the tips of fibres meeting at the anterior pole. We also describe a method for whole mount preparation, which allows observation of the polarized distribution of primary cilia at the apical surface of lens fibres. Given its relative simplicity, at least in cellular terms, and its requirement for a high degree of precision in cellular alignment and orientation, we predict that the lens will be an excellent model system to help elucidate the role of cilia and PCP components in the development of three-dimensional organization in tissues and organs.

Key words: Planar cell polarity (PCP), Mammalian eye lens, Lens fibre cell, Lens suture, Primary cilium, Whole mount

1. Introduction

Planar cell polarity (PCP) signaling provides a mechanism for aligning cells in a coordinated pattern along a particular axis to form functional tissues or organs. PCP was first identified in invertebrates and is now well established in several vertebrate systems. However, growing recognition of its importance in regulating many vital biological processes has led to an increased focus on PCP and its role in a wide range of tissues and organs (1–4). From studies of invertebrate model systems, several essential genes have been identified and most of these genes are now recognized as common players in vertebrate systems. Whilst there are many similarities between invertebrate and vertebrate systems that exhibit PCP, there are some differences. One such difference is the emerging recognition of the

Kursad Turksen (ed.), *Planar Cell Polarity: Methods and Protocols*, Methods in Molecular Biology, vol. 839,
DOI 10.1007/978-1-61779-510-7_12, © Springer Science+Business Media, LLC 2012

key role that primary cilia play in PCP in vertebrates (5–9). How cilia and PCP proteins cooperate to coordinate precise cell alignment and orientation has become a major focus of research.

Recently, we reported that the eye lens exhibits PCP, with each lens fibre cell having an apically situated cilium and in most cases this is polarized towards the anterior pole (Fig. 1; 10). Concentrically arranged fibres are precisely aligned as they elongate along the anterior–posterior axis and orientate towards lens poles where they meet equivalent fibres from other segments to form characteristic sutures. This global alignment is regulated by the PCP pathway. Because of its lack of cellular complexity and its distinctive polarity, the lens has many advantages for PCP studies. Moreover, the ease with which lens can be isolated from other eye tissues as well as the ability to prepare pure populations of the two forms of lens cells make it a very convenient tissue for study. Thus, we predict the lens will be a valuable model system to help elucidate the role of cilia and PCP components in development of three-dimensional organization in tissues and organs.

We have developed several new techniques to observe lens PCP in detail. However, the standard histological approach of using paraffin-embedded lenses or whole eyes to study lens cell

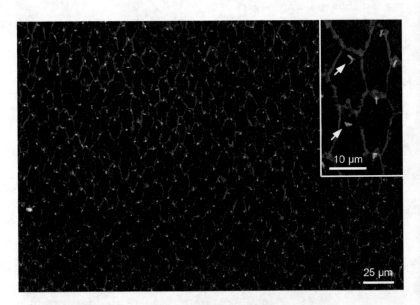

Fig. 1. Hexagonally aligned lens fiber apical tips with polarized centrosomes/primary cilia. A whole mount was prepared from a P34 rat lens and immuno-stained with anti-pericentrin and anti-β-catenin antibodies. Pericentrin reactivity localizes the centrosome (and associated primary cilium; see ref. 10) and β-catenin reactivity demarcates the cell borders and shows the regular packing of fiber cells as viewed in cross section. Virtually all cells have a centrosome/primary cilium that is polarized to the side of the cell that faces the anterior pole (top of page). The polarization of the centrosome/primary cilium (*arrows*) in fibres and their hexagonal shape is clearly evident at higher magnification (inset).

alignment and orientation also gives valuable information. We mostly use rats and mice as experimental material is readily available and, particularly in the case of mice, transgenic models and mutants can be generated or accessed from other researchers. In PCP mutants like Loop-tail (Lp) or secreted frizzled-related protein 2 (Sfrp2) overexpressing mice, lenses do not develop the spheroidal shape typical of their wild-type littermates but tend towards a flatter overall appearance (in Lp lenses) or become flatter anteriorly and more pointed posteriorly (in Sfrp2 mice) (10, 11). Cross sections that are horizontal to the lens equator also give a lot of information about cellular organization. From this perspective it is clearly evident that the normal lens consists of layers of well-organized, concentrically arranged fibres with similar global alignment and orientation. Any disturbance of this can be readily assessed in experimental PCP models. For example, in Sfrp2 overexpressing mice, although fibres show similar alignment within groups, different groups usually exhibit different orientations and this results in the formation of distinct boundaries where different groups of fibres meet. Because paraffin embedding and subsequent sectioning methodologies involve standard histological procedures we will not deal with these here, rather we will concentrate on the new techniques that we have developed to study PCP in the lens.

Here we describe two advanced methods for examining lens PCP in detail. First we introduce a method to observe the anterior suture pattern by using whole lens staining. This analysis provides information on whether global PCP is maintained in the particular lens under investigation. Second, we describe a method to visualize the primary cilium at the fibre apical tip to determine if it shows characteristic polarized distribution as well as to assess integrity of local PCP.

2. Materials

2.1. Samples

1. Mice: Transgenic mice (e.g., Sfrp2 overexpressing) or mutant or knockout mice (e.g., Lp, Crash) aged between embryonic day (E) 16.5 to about 1-month postnatal.

2. Rats: Wistars (about 30 days old). Cellular alignments are similar in the lenses of mice and rats, but the latter provide larger-sized lenses that can often facilitate assessment of normal cellular patterns.

2.2. Dissection Tools

1. Standard forceps.

2. Scissors with curved tips.

3. 35 mm cell culture dishes (Nunc).

4. 10 mL tubes with screw caps.

5. Dissection microscope and light source (Leica).

6. Two pairs of fine forceps (Dumont Jeweller's Forceps #5).

7. Filter papers (3 M, Wattman).

2.3. Solutions

1. Sterile essential medium (M199) with Earle's salts (Invitrogen) supplemented with 0.1% bovine serum albumin (Sigma-Aldrich), 2 mM L-glutamine, 50 U/mL penicillin, 50 µg/mL streptomycin and 2.5 µg/mL of fungizone.

2. PBS: Prepared from tablets (Roche).

3. 100% Methanol.

4. 4% (w/v) paraformaldehyde in PBS.

5. Dye dilution buffer: 1% NP-40 in PBS.

6. Fluorescent dyes: TRITC-lectin (Sigma), Alexa 488-phalloidin (Invitrogen).

7. 2% (w/v) agarose in PBS; prepare in 100 mL screw capped glass bottle and melt in microwave oven.

8. Washing buffer: 0.1% (w/v) BSA in PBS.

9. Blocking solution: 10% normal donkey serum (Millipore) in washing buffer.

10. Primary antibody solution: rabbit anti-pericentrin antibody (Abcam) and mouse anti-β-catenin antibody (BD) with 1.5% normal donkey serum in washing buffer.

11. Secondary antibody solution: Alexa 488-conjugated donkey anti-rabbit IgG and Alexa 594-conjugated donkey anti-mouse IgG (Invitrogen) in washing buffer.

2.4. Other Materials

1. Water repelling marker: Dako Pen (Dako).

2. Humid chamber (Plastic airtight container with damp papers and supports for slides).

3. Fine pliers.

4. Water-based mounting medium: Aqua-Poly/Mount (Polysciences), Pristine Mount (Invitrogen).

5. 22 mm diameter coverslips.

6. Glass slides.

7. Plastic transfer pipette.

2.5. Microscope

1. Upright microscrope (Axioskop2, Zeiss) with laser confocal system (LSM5 Pascal, Zeiss).

3. Methods

In mice and rats, PCP becomes progressively more prominent in lens fibre cells during postnatal development; consequently their highly organized alignment and orientation can be best visualized in adults rather than in neonates. Lenses also continue to grow throughout life, so for ease of manipulation it may be beneficial to use adult lenses. If the gene of interest causes embryonic lethality and it is not possible to get postnatal samples, aim to get the oldest stage samples as possible, ideally at E18.5, since by this time, suture alignment has been established to some extent and it is possible to determine if suture formation is defective. However, visualizing the positioning of cilia at the apical tips of fibres may be not possible even at postnatal day 3 because of the inherent technical difficulties of preparing whole mounts from small lenses. Conditional knockouts overcome such developmental problems as well as size issues. The availability of lens-specific Cre recombinase transgenic mice provides a practical solution to these problems. So far, several lines of lens-specific Cre mice that have variable Cre expression timings and patterns are available (12, 13). Usually lens defects do not affect viability of mice and thus these models permit access to older-stage lenses. The other major advantage of conditional knockouts is that the phenotype in the tissue of interest is unlikely to be due to secondary effects resulting from damage to an adjacent tissue.

3.1. Preparation of Lens Samples

1. Eyeballs are removed, by external approach, from euthanised animals and transferred to CO_2 equilibrated, pre-warmed M199 medium in a 35 mm culture dish. For adult animals, use curved scissors and press down alongside two edges of the eye with the tips of scissors so that the eyeball protrudes; then pinch it off by sliding the scissor tips underneath eyeball. In younger animals, before about postnatal day 13, eyes are still closed and eyelids need to be removed. For this, lift eyelid with forceps and cut it off with scissors to expose eyeball underneath, then proceed as described for older animals.

2. To dissect out lenses from eyeballs, tear the posterior part of the eyeball by inserting the tips of fine forceps (use a pair of forceps in each hand) into the hole left by the cut optic nerve. Carefully open eyeball by tearing the retina layer and removing surrounding ocular tissue from lens. Filter paper can be used to wipe the tips of the forceps as these get sticky with tissue debris. Discard removed tissues from culture medium.

3.2. Anterior Suture Observation

1. Transfer dissected eyeballs to a 10 mL tube and fix with 4% PFA/PBS overnight at room temperature. After 1 h wash with PBS, isolate lenses by following Subheading 3.1, step 2. Tissues

may become more elastic and opaque after fixation, but it is still possible to isolate lenses from eyeballs (see Notes 1 and 2).

2. Transfer isolated lenses to TRITC-lectin or Alexa 488-phalloidin (or both) in dye dilution buffer and incubate for a minimum of 2 days at room temperature with gentle agitation. To transfer small lenses without damaging them, use a plastic transfer pipette and take lenses up into the tip of the pipette with a small amount of solution.

3. Wash lenses in PBS for 1 h at room temperature.

4. Transfer a lens to a coverslip with anterior pole side down (i.e., epithelium side facing the coverslip). Remove as much excess PBS as possible. Drop melted 2% agarose/PBS onto the lens and leave until agarose sets and lens is fixed in place (see Note 3).

5. To observe under the microscope, invert the coverslip over a glass slide and rest on two platforms, 3 mm tall and separated by about 15 mm, so that the lens is suspended in the space between the platforms. The platforms can be made by breaking a glass slide into six strips, each 1 mm thick, and forming two separate platforms each made up of three glass strips glued onto the slide.

6. Under the microscope, the lens capsule and lens epithelial layer will be visualized first (Fig. 2a). Apical tips of lens fibres are visible just beneath lens epithelial cells (Fig. 2b). Collect a confocal slice image that contains a suture image (Fig. 2c, d) or reconstitute a 3D projection image to show suture structure in detail (Fig. 2f) (see Note 4).

3.3. Whole Mount Preparation

In this method, apical tips of fibres are preserved as an attachment on the epithelial layer when lens capsule is peeled off from the fibre mass (Fig. 3a). The extent of fibre remnants maintaining attachment to the epithelium may vary between samples; sometimes almost the whole epithelial sheet is covered by fibre tips so that the complete suture structure may be visible (Fig. 3b). In other cases, only a few small patches of fibre tips may be retained on the epithelial layer. When this happens, try changing the peeling speed and/or the angle at which the capsule is pulled away from the fibre mass. To distinguish epithelial cells and fibre tips, E-cadherin staining may be helpful since E-cadherin is detected only in epithelial cells (Fig. 3b). Apical tips of fibres have a characteristic hexagonal shape whereas the epithelial cells have random cobblestone-like appearance, so this morphological difference may also help to distinguish these two types of cells. This is a modified method originally developed to prepare lens epithelial explant cultures (14).

1. Fix isolated lenses with 100% methanol for 45 s and wash twice with PBS for 5 min each. Keep lens-containing dishes on ice. Fixed lenses can be stored in PBS at 4°C for up to 1 week.

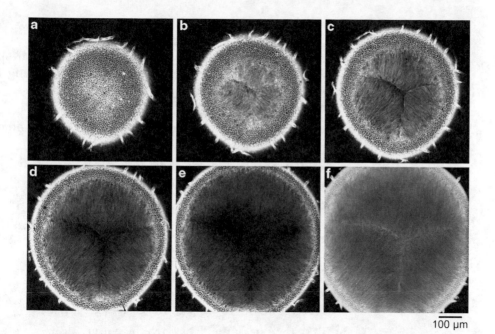

Fig. 2. Postnatal 5-day mouse lens stained with TRICT-lectin for 2 days and sequentially imaged by confocal microscopy starting from the anterior pole (×20 objective lens, 2 μm optical sections). (a) In the first slice, cobblestone-like lens epithelial cells are in focus. Blood vessels that are associated with the lens surface are also evident at the periphery. (b) The focal plane has now passed through the central lens epithelium region and apical tips of fibres are visible. (c, d) In these planes, the typical Y-shaped suture is clearly recognized. (e) Deeper within the lens, fibres along the suture are devoid of staining because of limited dye penetration. (f) Three-dimensional projection image reconstituted from a series of sliced images using the projection tool of Zeiss LSM software.

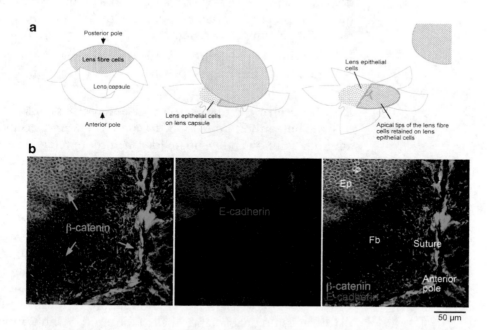

Fig. 3. Preparation of lens whole mounts. (a) Diagram showing how the lens capsule is peeled away from the lens fibre cells. Lens epithelial cells remain firmly attached to the lens capsule; however, when the fibres are removed, in some regions, remnants of their apical ends are retained on the apical surface of the epithelial cells. (b) β-catenin is present at the margins of both epithelial and fibre cells but E-cadherin is specific for epithelial cells. In this whole mount, a large region near the anterior pole of the lens retains remnants of the apical ends of lens fibre cells. Abbreviations: *Ep* epithelial cells, *Fb* fibre cells. Adapted from ref. 10.

2. Transfer one or two lenses into a new cell culture dish filled with PBS. Place the lens anterior pole side down (facing the culture dish surface). By using two fine forceps, make a small tear on the capsule at the posterior pole.

3. From the small tear, make five or six tears toward anterior surface until just passing the equatorial region.

4. Peel off the capsule along the tears and expose posterior part of lens fibres. Remove lens fibre mass.

5. Gently press a tip of one flap of the capsule onto the plastic culture dish until the capsule is immobilized and sticks to the culture dish surface. Make one more attachment to fix the capsule onto the dish.

6. Stretch the lens capsule slightly to make it as flat as possible. If the capsule is rolled up when the fibre mass is removed try to unravel it before fixing it to the dish (see Note 5).

7. Pin all of the tips of the flaps onto the culture dish.

3.4. Immunostaining of Whole Mount

1. Decant PBS and remove remaining moisture around whole mounts with tissue paper. Circle whole mounts with Dako pen to make water-repellent barrier. Add PBS immediately into the circle to prevent any tissue dehydration.

2. Replace PBS with blocking medium containing 10% normal donkey serum. Keep whole mount-containing dishes in a humid chamber to prevent evaporation. Incubate for 30 min at room temperature or overnight at 4°C (see Note 6).

3. Remove blocking solution and apply primary antibody solution. Incubate samples for 1 h at 37°C or overnight at 4°C in humid chamber.

4. Remove primary antibody solution and wash samples with washing buffer three times for 5, 10 and 15 min. Add secondary antibody solution and incubate for 1 h at room temperature in the dark (place in box or some other container that blocks out the light; see Note 7).

5. Remove secondary antibody solution and wash samples with washing buffer three times for 5, 10 and 15 min. Keep samples in dark.

6. Break culture dish wall using pliers and remove surrounding wall completely to convert the dish into a disc. Add a drop of water-based mounting medium onto the whole mount and seal with a round coverslip. The disc of culture dish can then be placed in its original lid and labelled for future identification.

7. Place the disc containing the whole mount on a glass slide with a drop of water or glycerol to act as a glue to hold the specimen in place. Observe fibre apical tips with cilia using a 40 times objective lens or higher.

4. Notes

1. For whole lens staining, an alternative method that can be applied to large adult lenses is to dissect lenses prior to fixation. In this case treat dissected lenses with 4% PFA/PBS for 1 h before washes. 100% methanol can be used for fixation by treating lenses for 45 s. Note phalloidin does not work with alcohol-based fixatives, so that only lectin staining is applicable if the lenses are fixed with methanol.

2. 10% Neutralized formalin usually contains methanol to promote stability; however, be aware that this may affect phalloidin staining.

3. The anterior pole should be centrally positioned but sometimes this can be difficult to ascertain. If you can see the posterior suture, this helps to centre the lens, but otherwise try to use iris remnants on the lens as an indicator of lens orientation.

4. Efficiency of dye penetration is enough to observe the surface suture but usually, dye penetration is limited to the cortical region of lenses (see Fig. 2e. In this sample, lectin still stains epithelial cells and outer regions of fibres, but the central portion of fibres is devoid of staining). Longer incubation times may enhance intensity of signal and depth of dye penetration, but it appears that after 2 days incubation no further penetration is achieved.

5. Whole mounts should be prepared so that they are flat, otherwise wrinkles/folds interfere with microscopy.

6. We recommend normal donkey serum as blocking reagent of immunostaining. Normal goat serum tends to give high background signal especially at the centrosome/cilium. Use freshly thawed serum and keep it at 4°C as repeated freeze-thaw causes higher background signals.

7. Nucleus detecting dyes (Hoechst, DAPI etc.) can be included in secondary antibody solution.

Acknowledgements

This work was supported by NHMRC (Australia), NIH (USA, R01 EY03177), ORIA, Australia and The Sydney Foundation for Medical Research. Y.S. was supported by an Endeavour Fellowship, Australia and The Sydney Eye Hospital Foundation. Some research illustrated here was undertaken as part of the Vision CRC, New South Wales, Sydney, Australia.

References

1. Lawrence PA, Struhl G, Casal J (2007) Planar cell polarity: one or two pathways. *Nat Rev Genet* 8:555–563.

2. Seifert JR, Mlodzik M (2007) Frizzled/PCP signaling: a conserved mechanism regulating cell polarity and directed motility. *Nat Rev Genet* 8:126–138.

3. Strutt D (2008) The planar polarity pathway. *Curr Biol* 18:R898–902.

4. Wang Y, Nathans J (2007) Tissue/planar cell polarity in vertebrates: new insights and new questions. *Development* 134:647–658.

5. Axelrod JD (2008) Basal bodies, kinocilia and planar cell polarity. *Nat Genet* 40:10–11.

6. Park TJ, Haigo SL, Wallingford JB (2006) Ciliogenesis defects in embryos lacking inturned or fuzzy function are associated with failure of planar cell polarity and Hedgehog signaling. *Nat Genet* 38:303–311.

7. Ross AJ, May-Simera H, Eichers ER, Kai M, Hill J, Jagger DJ, Leitch CC, Chapple JP, Munro PM, Fisher S, Tan PL, Phillips HM, Leroux MR, Henderson DJ, Murdoch JN, Copp AJ, Eliot MM, Lupski JR, Kemp DT, Dolfus H, Tada M, Katsanis N, Forge A, Beales PL (2005) Disruption of Bardet–Biedl syndrome ciliary proteins perturbs planar cell polarity in vertebrates. *Nat Genet* 37:1135–1140.

8. Simons M, Walz G (2006) Polycystic kidney disease: cell division without a c(l)ue? *Kidney Int* 70:854–864.

9. Singla V, Reiter JF (2006) The primary cilium as the cell's antenna: signaling at a sensory organelle. *Science* 313:629–633.

10. Sugiyama Y, Stump RJ, Nguyen A, Wen L, Chen Y, Wang Y, Murdoch JN, Lovicu FJ, McAvoy JW (2010) Secreted frizzled-related protein disrupts PCP in eye lens fiber cells that have polarized primary cilia. *Dev Biol* 338, 193–201.

11. Chen Y, Stump RJ, Lovicu FJ, Shimono A, McAvoy JW (2008) Wnt signaling is required for organization of the lens fiber cell cytoskeleton and development of lens three-dimensional architecture. *Dev Biol* 324:161–176.

12. Ashery-Padan R, Marquardt T, Zhou, X, Gruss P (2000) Pax6 activity in the lens primordium is required for lens formation and for correct placement of a single retina in the eye. *Genes Dev* 14:2701–2711.

13. Zhao H, Yang Y, Rizo CM, Overbeek PA, Robinson ML (2004) Insertion of a Pax6 consensus binding site into the alphaA-crystallin promoter acts as a lens epithelial cell enhancer in transgenic mice. *Invest Ophthalmol Vis Sci* 45:1930–1939.

14. Lovicu FJ, McAvoy JW (2008) Epithelial explants and their application to study developmental processes in the lens. In: Tsonis PA (ed) Animal Models in Eye Research, Academic Press, New York.

Chapter 13

Examining Planar Cell Polarity in the Mammalian Cochlea

Helen May-Simera and Matthew W. Kelley

Abstract

The mammalian cochlea offers a unique opportunity to study the effects of planar cell polarity signaling during vertebrate development. First, convergence and extension play a role in outgrowth and cellular patterning within the duct, and second, hair cell stereociliary bundles are uniformly oriented towards the lateral edge of the duct. Defects in convergence and extension are manifested as a shortening of the cochlea duct and/or changes in cellular patterning, which can be quantified following dissection from mouse mutants or observed directly using an in vitro outgrowth assay. Changes in stereociliary bundle orientation can be observed and quantitated using either fluorescent tags or scanning electron microscopy (SEM) to visualize individual bundles. The high degree of regularity in many aspects of cochlear anatomy, including cellular patterning and stereociliary bundle orientation, makes it possible to detect subtle changes in the development of PCP in response to either genetic or molecular perturbations.

Key words: Cochlea, Hair cells, Development, Organ of Corti, Stereociliary bundle, Kinocilium, Microtubules, Actin, Vangl2, Wnt

1. Introduction

In mammals, auditory signals are detected in the cochlea, a coiled duct that extends ventrally from the vestibular portion of the inner ear. All aspects of the inner ear, including the cochlea, are encased in bone (the otic capsule or bony labyrinth), which is itself embedded in the temporal bone. The dorsal surface of the cochlear duct (referred to as the floor) contains the organ of Corti, the specialized sensory epithelium that converts sound-generated pressure waves into neural signals. The organ of Corti is comprised of specialized epithelial cells, including both mechanosensory hair cells and non-sensory supporting cells, which are organized in a highly ordered and regular pattern that extends from the base of the cochlea to its apex. The organ of Corti comprises two distinct types of mechanosensory

Kursad Turksen (ed.), *Planar Cell Polarity: Methods and Protocols*, Methods in Molecular Biology, vol. 839,
DOI 10.1007/978-1-61779-510-7_13, © Springer Science+Business Media, LLC 2012

hair cells, inner hair cells and outer hair cells. The two different hair cell types were initially identified based on positional and morphological differences, but subsequent studies have demonstrated physiological and genetic distinctions as well (Fig. 1a). The organ of Corti also contains several different types of non-sensory cell types, collectively referred to as supporting cells, which surround the hair cells and serve auxiliary, but no less important, roles in auditory function (Fig. 1a).

Mechanosensory hair cells are so named because of the presence of a cluster of modified microvilli referred to as a stereociliary bundle located on the luminal surface of the hair cell. Hair cells are found in the inner ears of all vertebrates as well as in the lateral line neuromasts of fish and other aquatic vertebrates. In a scanning electron micrograph, stereociliary bundles appear morphologically similar to a tuft of hair (Fig. 1b). These bundles act as the primary transducers of mechanosensation. Sound- or motion-induced pressure waves cause the bundle to pivot at its base, leading to an influx of positively charged ions and cellular depolarization (Fig. 1d) (1). All stereociliary bundles are arranged in a staircase pattern with the height of the individual stereocilia in each row increasing progressively across the bundle (2). Stereociliary bundles are directionally sensitive such that a deflection towards the tallest row of stereocilia leads to cellular depolarization, while deflection away from the tallest row hyperpolarizes the cell. Deflections perpendicular to this axis do not lead to a change in the resting state of the cell (1).

In the organ of Corti, the morphology of the stereociliary bundles is modified such that the bundles appear similar to a ")" or "W" shape (Fig. 1a–c). Inner hair cells usually have two main rows of stereocilia arranged in a ")" shape, while outer hair cells have three rows of stereocilia and the bundle has a notch on one side leading to a "W" morphology (3). In addition, the stereociliary bundles on all hair cells in the organ of Corti are oriented such that the tallest row of stereocilia is located at the lateral edge of the cell. This orientation is required for normal function because all sound-induced activity within the cochlear duct leads to deflections that are oriented exclusively along the medial-to-lateral axis (1).

Stereocilia are microvilli rather than true cilia and contain a dense actin core. However all stereociliary bundles also contain a single true cilium, referred to as a kinocilium, which is located adjacent to the tallest row of stereocilia (4, 5). While kinocilia are only present transiently on hair cells in the mammalian auditory system and degenerate in mice just prior to the onset of hearing, they persist throughout life in all other vertebrate hair cell types, including mammalian vestibular hair cells. The functions of kinocilia remain unclear. They are not required for mechanotransduction (6) and are clearly dispensable, at least in the mammalian auditory system, for all hair cell functions. Based on these observations, it has been suggested that kinocilia may play a role in the development and orientation of the stereociliary bundle.

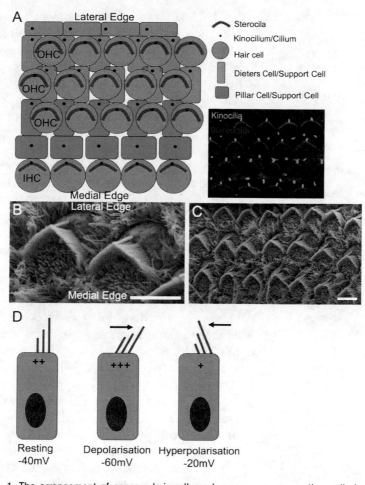

Fig. 1. The arrangement of sensory hair cells and non-sensory supporting cells in the mammalian cochlea. (**a**) Schematic representation of a whole-mount of the developing organ of Corti at P0 in a mouse. Stereociliary bundles (*red*) are located on the apical surface of hair cells (*orange*). Primary cilia (*black*) are located on all cells, at the lumenal surface. Stereociliary bundles appear as a flattened ")" shape on both inner and outer hair cells, although outer hair cell bundles will develop a "W" shape with continued maturation. Lateral and medial edges relative to the cochlear spiral are indicated. Micrograph illustrates labeling of stereocilia (*red*) and kinocilia (*green*) in a biological sample. (**b, c**) SEM images of mammalian outer hair cells. Mouse embryonic day 17.5, basal turn Scale bars, 5 μM. (**b**) High magnification view of two stereociliary bundles. Note the position of the single kinocilium at the vertex of the bundle. (**c**) Lower-magnification image of the three rows of outer hair cells. Note the uniform orientation of the stereociliary bundles towards the lateral edge of the epithelium. (**d**) Schematic illustrating effects of stereociliary bundle deflections on the resting state of a hair cell. Deflection of the stereociliary bundle towards the tallest row of stereocilia leads to opening of transduction channels, an influx of positively charged ions and a subsequent depolarization of the cell. In contrast, deflection towards the shortest row of stereocilia leads to closing of transduction channels and hyperpolarization of the cell.

Each kinociliam develops from a single primary ciliam located at the luminal surface of each developing hair cell. Initially located in the center of the lumenal surface, the developing kinociliam gradually migrate towards the lateral edge (5, 7–10). As this happens, microvilli elongate adjacent to the migrating cilium to develop as stereocilia. Mutations that lead to defects in different aspects of cilium formation or migration often lead to defects in bundle formation or orientation, suggesting that the developing kinocilia play a key role in bundle development (11, 12)

In mouse, developing auditory hair cells can first be discerned between embryonic days 13 and 15 (E13-E15) (13). As these cells differentiate from surrounding progenitor cells they undergo a period of migration and patterning that leads to their final ordered arrangement by approximately post-natal day (P3). Between E15 and E16, developing kinocilia begin their migration towards the lateral edge of the lumenal hair cell surface and adjacent microvilli elongate to form stereocilia. While the initial migration of each developing kinocilium is directed towards the lateral edge of the hair cell, many developing bundles are not appropriately oriented when this migration is completed. However, a subsequent period of refinement occurs in which developing bundles gradually reorient to ultimately achieve uniform orientation by approximately P10 (14).

The processes of cochlear outgrowth and stereociliary bundle orientation are both controlled through the PCP pathway. Examples of planar cell polarity (uniform orientation within an epithelial plane) have been described in animals as diverse as Drosophila and mammals and in virtually all cases, the PCP pathway has been shown to regulate this process. Moreover, in vertebrates, the PCP pathway has also been shown to regulate additional developmental processes such as the coordinated remodeling of epithelial sheets through convergence and extension (15).

The PCP pathway utilizes a core set of structurally related and conserved proteins that include transmembrane proteins such as Vangl1 and 2 and members of the Frizzled family. In addition, in vertebrates the pathway also requires Wnt molecules, such as Wnt5a and Wnt11 (16). However, a similar role for Wg the invertebrate ortholog of Wnt, has not been demonstrated in *Drosophila*. Mutations in mammalian PCP genes, such as *Vangl2*, *Scrb1*, *PTK7/CCk-4*, *Celsr1*, and *Dvl1,2* lead to shortened cochlea ducts, defects in the patterning of hair cells and supporting cells, misoriented stereociliary bundles, and hearing defects (14, 17–19).

The following protocols describe analysis of cochlea extension and stereociliary bundle orientation, analysis of bundle morphology by scanning electron microscopy (SEM), and cochlea outgrowth assays performed in embryonic cultures.

2. Materials

All reagents were of AnalaR grade. Solutions were made using molecular grade distilled and deionized water unless specified.

2.1. Reagents for Dissection and Culturing

1. 1× Phosphate buffered saline (Gibco) (PBS).
2. 70% Ethanol.
3. Dissection medium: 1× Hanks' Balanced Salt Solution (Gibco) (HBSS), 5 mM Hepes (Gibco), pH adjusted to 7.2–7.3 and filter sterilized. Stored at 4°C for up to 1 month.
4. Culture medium: Dulbecco's Modified Eagle Medium (Invitrogen) (DMEM) supplemented with 10% fetal bovine serum (Gibco), 1× N2 supplement (Invitrogen) and 0.01 mg/ml ciprofloxin. Made under sterile conditions and stored at 4°C for up to 1 week.
5. Matrigel (BD Biosciences).
6. Mattek Dishes.
7. OS-30 Solvent (Dow-Corning).

2.2. Reagents for Immunohisto-chemistry

1. 1× Phosphate buffered saline (PBS).
2. Paraformaldehyde (Electron Microscopy Sciences) (PFA): prepare a 4% solution in 1× PBS made fresh each time.
3. 0.1 M Tris-HCl (pH 7.5), 0.15 M NaCl, 0.1% Triton X-100 (Triton buffer).
4. 10% Goat Serum in Triton buffer (Triton block).
5. AffiniPure Fab Fragment Donkey Anti-Mouse IgG (H+L) (Jackson ImmunoResearch): added to Triton buffer (1:200) to make Triton block for blocking step when using antibodies raised in mouse.
6. Primary antibodies and cellular markers: Monoclonal anti-acetylated-alpha-tubulin (611-B1, Sigma, 1:800), Monoclonal anti-γ-tubulin (GTU-88, Sigma, 1:200), anti-Myosin 6 or anti-Myosin 7a (Proteus Biosciences, 1,000).

   Secondary antibodies: Anti-rabbit or anti-mouse conjugated with Alexa Fluor 488 or 594 (Invitrogen; 1:200).

   Alexa Fluor 488 or 546-conjugated phalloidin (Invitrogen).
7. Fluoromount-G (SouthernBiotech).

2.3. Reagents for Scanning Electron Microscopy

1. 1× Hanks' Balanced Salt Solution (Gibco) (HBSS).
2. 0.1 M Hepes (Gibco) in HBSS (Hepes buffer).
3. 2.5% Glutaraldehyde Grade 1 (Sigma-Aldrich), 4% paraformaldehyde EM Grade (Electron Microscopy Sciences), 10 mM $CaCl_2$ (Quality Biological) in Hepes buffer (EM Fix).

4. 1% Osmium tetroxide (OsO_4) in Hepes buffer or water: Diluted from 4% Osmium tretraoxide Solution (Fluka Analytical).

5. 1% (w/v) Tannic acid (Sigma-Aldrich) in water.

6. Graded ethanol solution series: 30, 50, 70, 90, 95% prepared with 200 proof ethanol and distilled water.

7. 100% Ethanol 200 proof (see Note 1).

2.4. Tools/Equipment

1. Black sylgard dish (see Note 2).

2. Scissors.

3. Fine forceps (no. 5) and blunt forceps.

4. Dissecting microscope.

5. Microscope slides.

6. Microscope cover slips (22 mm × 40 mm × 0.15 mm).

7. Transfer pipettes.

8. Minutien pins (Fine Science Tools).

9. 60- or 100-mm culture dishes (MatTek).

10. 35-mm tissue culture dishes, No. 0 coverslips (MatTek).

11. 37°C Incubator.

12. Basket for SEM preps.

13. Scanning electron microscopy studs (TED PELLA).

14. PELCO Tabs: Carbon adhesive (TED PELLA).

3. Methods

3.1. Cochlea Dissection

3.1.1. Dissection of the Temporal Bones

1. Animals can be euthanized at any age between E13 and P3 (see Note 3). Dissection is possible at older ages but becomes progressively more difficult because of ossification of the bony labyrinth and temporal bone. The head is removed to allow for cranial dissection.

2. A mid-line sagittal incision is made beginning at the nose and extending caudally. The brain is removed from each half of the skull.

3. With the brain removed, the temporal bones, which contain the developing bony (cartilaginous at this point) labyrinths of the inner ear, can be visualized and the bony labyrinths can be isolated from the skull (Fig. 2a).

4. The dissected bony labyrinths can be fixed (see below) before further dissection of the cochlea duct (Fig. 2b).

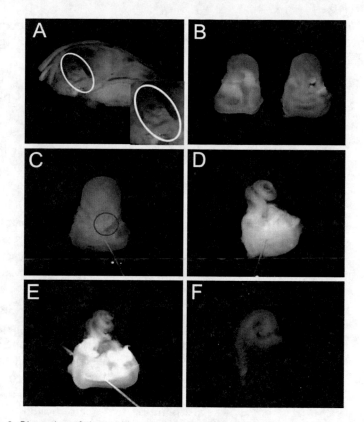

Fig. 2. Dissection of the cochlea. (**a**) Mid-line saggital dissection of P0 mouse pup with brain removed. Position of the bony labyrinth is circled. Inset, high-magnification view of the bony labyrinth. (**b**) Dorsal (*left*) and ventral (*right*) views of dissected bony labyrinths. The cochlea is located towards the *top* of each image. (**c**) A bony labyrinth that has been positioned for further dissection. A single pin has been placed through the vestibular system to aid in further dissection. *Red circle* indicates position of the oval window. (**d**) A bony labyrinth as in (**c**), but after removal of the outer cartilage of the cochlea, exposing the cochlear duct. (**e**) Same view as in (**d**), but after removal of the developing Reisner's membrane, exposing the floor of the cochlear duct, including the sensory epithelium. (**f**) The cochlear duct, isolated from the vestibular region of the labyrinth and ready for mounting

3.2. Fixation

3.2.1. Fixation of the Bony Labyrinth/Cochlear Duct

1. For regular immunohistochemistry, bony labyrinths are fixed in 4% PFA overnight at 4°C on a Nutator. Fixation can also be done for 2 h at room temperature also with nutation. Because of the sensitivity of certain antibodies, variations in the length and composition of fixation may be required. Following fixation, tissue is washed three times for 5 min in PBS. Samples can then be stored for up to 2–3 weeks in PBS at 4°C prior to processing.

2. For scanning electron microscopy analysis, temporal bones are fixed in SEM fix for 2 h at room temperature on a Nutator. Fixation is followed by three washes of 5 min each with Hepes buffer and can then be stored until further processing in Hepes buffer at 4°C for a few days (see Note 4).

**3.3. Dissection
of the Cochlea Duct**

1. Further dissection of the bony labyrinths to expose the cochlea epithelium is done in a black Sylgard-coated (see Note 2) dissecting dish containing PBS.

2. The use of minutien pins to immobilize the tissue can be helpful for fine dissection. Pins should be placed through the vestibular portion of the bony labyrinth with the ventral aspect of the cochlear spiral facing upwards (Fig. 2c).

3. Using fine forceps and starting at the oval window (Fig. 2c), the outer cartilage of the cochlear spiral is removed to expose the cochlea duct (Fig. 2d).

4. Once the duct is exposed, the developing Reissner's membrane (the ventral surface of the cochlear duct) must be removed to expose the sensory epithelium located on the dorsal aspect of the duct (Fig. 2e). This is most easily done by pinching the Reissner's membrane at the base of the cochlea duct and then gently prying the membrane off in an upward motion.

5. Once this is completed, the dorsal aspect of the duct, which includes the organ of Corti, is exposed (Fig. 1e). A second membrane, the tectorial membrane is located on the surface of the organ of Corti. The tectorial membrane is optically clear and does not need to be removed for immunohistochemistry but should be removed for scanning electron microscopy. This membrane can be removed by peeling it off the surface of the organ of Corti starting at the base of the cochlea and peeling towards the apex. Once this is completed the tissue is ready for processing. The vestibular region is retained to aid in further preparation of the samples.

**3.4. Immunohisto-
chemistry**

1. Tissue is permeabilized by washing three times for 5 min with Triton buffer at room temperature on a nutator. Following permeabilization, tissue is blocked for a minimum of 1 h in triton block.

2. Incubation with primary antibodies diluted in triton block solution is ideally done overnight at 4°C, or at room temperature for 2 h.

3. After incubation with primary antibody, tissue is washed three times for 15 min each in triton buffer on a nutator.

4. Primary antibodies are detected using AlexaFluor tagged secondary antibodies (Molecular Probes; 1:1,000). Incubation is done with secondary antibodies diluted in Triton block in the dark for 1 h at room temperature.

5. Following three final washes in triton buffer on a nutator, tissue is mounted onto glass slides for imaging (Fig 3a).

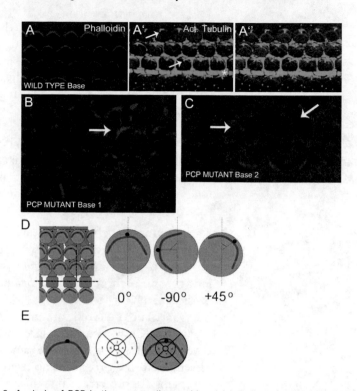

Fig. 3. Analysis of PCP in the mammalian cochlea. (**a**) Labeling of stereociliary bundles with fluorescently tagged phalloidin (*red*) illustrates the uniform orientation of stereociliary bundles in the basal turn of the cochlea in a P0 Wildtype mouse. (**a′**) Anti-acetylated-alpha-tubulin staining (*green*) labels the kinocilium on each hair cell (*arrows*) as well as other microtubule-rich structures such as at the apical poles of the pillar cells (*asterisk*). (**a″**) Merged image of (**a**) and (**a′**). Note the location of the kinocilia at the vertex of each stereociliary bundle. (**b–c**) Examples of misoriented stereociliary bundles in the basal turn of the cochlea in P0 PCP (*Vangl2*) mutants. *Arrows* indicated examples of mis-oriented bundles. Labeling as in (**a**). (**d**) Schematic representation of the criteria used to quantify the orientation of stereociliary bundles. The angle of rotation of the bundle is calculated relative to a line extending perpendicular to the row of pillar cells. (**e**) Schematic representation of the criteria for positional analysis of developing kinocilia. A segmented grid is laid over the circumference of the hair cell and the position of the kinocilium is noted.

3.5. Considerations for Immunohisto-chemistry

1. When using primary antibodies raised in mouse on mouse tissue, addition of AffiniPure Fab Fragment Donkey Anti-Mouse IgG (1:200, Jackson ImmunoResearch) to the blocking solution greatly decreases non-specific binding of anti-mouse IgG to mouse tissue.

2. For visualization of filamentous actin, including the actin-rich stereociliary bundles, tissue can be simultaneously stained with phalloidin conjugated to one of several different fluorophores (Invitrogen; 1:1,000) during the secondary incubation. In addition to the stereociliary bundle, phalloidin also marks other filamentous actin within each cell including the cortical actin located at the periphery of each epithelial cell.

3. While phalloidin is used to label the stereociliary bundles, an antibody against anti-acetylated-α-Tubulin is commonly used to stain the kinocilium. Similarly, anti-γ-Tubulin can be used as a specific marker for the basal bodies. For assessment of cochlea length, hair cells can be labeled using antibodies against Myosin 6 or Myosin 7a.

3.6. Mounting/Imaging of Fluorescently Labeled Cochleae

1. Following processing, cochlea can be further dissected to enable mounting of the entire cochlear spiral. First the sample is returned to a black Sylgard dish containing PBS and the vestibular region, underlying cartilage and mesenchyme are removed from the cochlear spiral (Fig. 2f).

2. The remaining cochlear spiral, which contains the inner sulcus, organ of Corti and outer sulcus is transferred into a drop of PBS on a microscope slide. The spiral is separated into pieces equivalent to one turn or less and oriented with the dorsal aspect of the spiral closer to the coverslip. This allows the full extent of the spiral to be visualized without overlap. The PBS is then wicked off with a Kimwipe and replaced with Fluoromount. Finally a cover slip is placed on top of the sample. The coverslip can be placed directly onto the drop with no need for any spacers. Number 0 coverslips are preferred, but thicker ones can also be used.

3. Imaging is carried out using either a standard epifluorescence microscope or a laser-scanning confocal microscope equipped with high magnification (60–100×), high numerical aperture (1.2–1.4) objectives.

3.7. Analysis

1. Length of the cochlear duct and sensory epithelium (organ of Corti). Determination of these values is straightforward. Once the duct has been labeled with phalloidin and a hair cell marker, the total lengths can be determined by simply measuring the entire length of the duct and the length of the sensory epithelium.

2. Stereociliary bundle orientation. Samples must be labeled with phalloidin (Fig 3a). Anti-acetylated tubulin or a hair cell marker as a counter stain may be useful but are not necessary. Because the organ of Corti develops in a gradient that extends from the mid-base towards both the apex and the base and development is not complete until approximately P14 in the mouse, it is important to compare regions at similar developmental stages. Therefore, the entire length of the sensory epithelium should be determined and specific positions along that length, such as 25, 50, and 75% from the base should be identified. However, it is important to consider that if the length of the cochlear duct is significantly shortened, then selection of positions based on percent from the base could result in inappropriate comparisons between experimental and control samples. In this case it

may make more sense to select positions based on absolute distance from the base. Regardless of the method used to determine measurement locations, once these locations are selected, images of phalloidin-labeled stereociliary bundles can be obtained (Fig. 3a–c). As discussed, it may also be useful to label kinocilia using an anti-acetylated tubulin antibody (Fig. 3a'). The orientation of each individual bundle can then be assessed by determining the rotation of the of bundle relative to a line extending perpendicular to the row of pillar cells that separates the inner and outer hair cells (Fig. 3d) (12, 14). Cells with a normal orientation are aligned along this perpendicular axis and so have a rotation of 0°. Angles of rotation should be determined using either a 360° notation or as absolute deviations from 0°. Orientation data should be collected from a minimum of 12 stereociliary bundles per sample from three separate samples. However, if a higher number of samples are examined then it may be possible to identify differences based on position along the duct or even between hair cells located in different rows at a single position.

3. Position of the kinocilium and basal body. Another aspect of PCP within the cochlea is the position of the developing kinocilia and associated basal body. To examine this, kinocilia and basal bodies can be labeled with anti-acetylated tubulin and anti-γ-tubulin respectively and their locations on the lumenal surface of the hair cells can be plotted by overlaying a positional grid and determining the distribution of cilia locations for a population of hair cells (Fig. 3e) (18). Again data should be collected from a minimum of 12 kinocilia per sample from three separate samples.

3.8. SEM Preparation

1. Following fixation in EM KX for 2 h at room temperature, temporal bones are dissected to expose the sensory epithelium; however, the vestibular portion of the bony labyrinth can be retained for ease of preparation.

2. The following steps are all carried out under ventilation (in a fume hood).

3. Samples are post-fixed in 1% OsO_4 in Hepes buffer for 1 h, followed by three washes with distilled water.

4. Samples are then incubated for 1 h at room temperature in each of the following solutions with three washes in distilled water in-between each step:

   1% Tannic acid freshly made in water and filtered prior to use.

   1% OsO_4 in water.

   1% Tannic acid in water.

   1% OsO_4 in water.

5. Samples are then dehydrated through a graded ethanol series by incubation for 10 min in each of the following dilutions: 30, 50, 70, 90, and 95%. Finally samples are transferred through three changes of 100% ethanol (from a newly opened bottle), 5 min for each change.

6. Following dehydration, the tissue is dried using a critical point drier at which point samples are ready to mount on conductive carbon adhesive on top of an SEM stud prior to imaging on an electron microscope.

7. Mounting is done under a stereomicroscope with extreme caution. Using a pair of fine forceps and a fine paint brush the samples are gently placed on the carbon adhesive with the cochlea spiraling upwards away from the SEM stud. Samples can be stored in a desiccator until imaging.

3.9. Extension Assay/ Culturing (Modified from Wang et al. (17))

3.9.1. Isolation of the Embryonic Inner Ear

1. All tools must be autoclaved in advance; procedures must be carried out in a sterile environment, preferably using a laminar flow bench.

2. Keeping tools and dissecting medium on ice blocks facilitates dissection.

3. Matrigel-coated MatTek culture dishes are made by diluting 150 μl of Matrigel in 3 ml of cold DMEM. Be sure to maintain the matrigel solution on ice at all times prior to addition to culture dishes. 100–200 μl of the diluted Matrigel is added to the center well of a 35-mm MatTek culture dish. The dishes are then placed in a 37°C incubator for at least 30 min to facilitate the precipitation of Matrigel onto the surface of the dish. Once made, these dishes can be stored at 37°C for up to 3 days prior to use. It is important not to place these dishes at 4°C as this will lead to the Matrigel going back into solution. Prior to adding tissue, Matrigel-coated culture dishes must be rinsed one time with DMEM or dissecting medium.

4. E13–E16 embryos are dissected from a timed-pregnant female. The uterus is removed and placed in a dish with cold dissection medium.

5. The embryos are removed and their heads detached. Heads are placed in a fresh culture dish with cold dissection medium.

6. The developing inner ears are then dissected as described above.

7. Once the inner ears have been isolated, they are transferred to a new culture dish containing cold dissection medium.

3.9.2. Dissection of the Embryonic Cochlea

1. For the further dissection of the cochlea the wider vestibular portion can be pinned to immobilize the tissue. Temporal bones are pinned so that the ventral side is facing upwards, with the apex spiraling up (Fig. 4a).

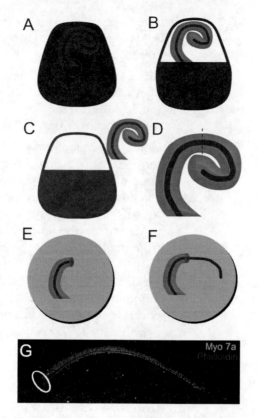

Fig. 4. Schematic illustration of the cochlear extension assay. (**a**) Isolation of embryonic bony labyrinth. Position of the oval window (*red circle*). (**b**) Similar view as in (**a**), but following exposure of the cochlear duct. The duct is retained intact and the location of the sensory epithelium is depicted in *red* for reference only. (**c**) Removal of the exposed cochlear duct from the underlying cartilage. (**d**) Position of the cut-site midway along the cochlea duct. (**e**) Plating of the basal portion of the cochlea duct. (**f**) Extension of the sensory epithelium from the cut site. (**g**) Immunostaining of extended culture. Hair-cells are labeled with anti-Myosin7a (*green*), and surrounding cells are labeled with phalloidin (*red*). The site of the original cut is circled.

2. The developing cartilage is removed using fine forceps (Fig. 4b). Inserting the bottom top of the forcep into the exposed oval window at the base of the cochlea spiral, one can snip upwards following the developing cochlea duct, being careful not to damage the underlying epithelium.

3. Following the spiral of the cochlea the cartilage can be removed very slowly ensuring that the ventral half of the cochlea duct is retained. For extension assays it is important to retain an intact cochlea duct.

4. Once the cochlear duct is exposed, it must be separated from the underlying cartilage (Fig. 4c).

5. Once isolated from all surrounding tissues, use fine forceps to snip the cochlear duct at the approximate halfway point along the basal to apical axis (Fig. 4d). The apical piece can be discarded as only the basal section will be plated.

3.9.3. Plating Basal Cochleae

1. The basal piece of the intact cochlea duct is transferred to a Matrigel-coated tissue culture dish containing 150 μl of culture medium.

2. The piece is then oriented with the ventral (roof) side of the duct facing upwards within the dish (Fig. 4e). This will place the dorsal side (the floor including the sensory epithleium) of the duct on the Matrigel-coated surface.

3. Cultures are placed in a 37°C incubator and can be maintained for 3–5 days. During this time, the epithelium comprising the floor of the duct, including the developing organ of Corti, extends outwards from the cut site (Fig. 4f, g). In contrast, the roof of the duct does not extend, and so can be used as a landmark to determine the amount of extension. No extension is seen from the basal end of the cochlea (see Note 5) (17).

3.9.4. Processing of Cochlea Extension

1. Following fixation with 4% PFA for 30 min at room temperature, immunostaining can be carried out to identify the extent of cochlear extension. Routinely a hair cell marker (anti-Myosin 6 or anti-Myosin 7a) is used for identification.

2. After immunostaining, coverslips can be removed from the culture dishes by soaking the bottom of the dish in OS-30 solvent for 45 min at room temperature.

3. Coverslips can be mounted onto microscope slides using Fluormount and imaged using an epifluorescence microscope.

4. Notes

1. For the final rinses with 100% ethanol, a freshly opened bottle of 200 proof ethanol should be used.

2. To make a black Sylgard dish: Sylgard base component is mixed with powdered charcoal to obtain an opaque black color. The curing agent is added and the sylgard is poured into one or more petri dishes. Petri dishes are dried under a vacuum to remove trapped air bubbles. Dishes must be completely dried (allow a few days) prior to use.

3. The most convenient age to use is P0; however, any age from E17 to adult can be used. Adult cochleae require decalcification prior to further dissection after fixation. For decalcification post-fixation, adult cochleae are stored in 4.13% EDTA, pH 7.3, in PBS at 4°C for 3–4 days under rotation. The solution is replaced daily and the tissue rinsed three times in PBS for 5 min on the final day.

4. Storing dissected samples for scanning electron microscopy in Hepes buffer for longer than a couple of days is not recommended even at 4°C.

5. If the base of the cochlear duct is damaged during the dissection, it can be difficult to identify the apical side of the cochlear duct in explants after many days in vitro. Therefore, it is crucial to maintain tissue integrity when isolating the cochlear duct from the underlying tissue/mesenchyme.

References

1. Hudspeth, A.J., and Corey, D.P. (1977). Sensitivity, polarity, and conductance change in the response of vertebrate hair cells to controlled mechanical stimuli. *Proc Natl Acad Sci USA 74*, 2407–2411.

2. Frolenkov, G.I., Belyantseva, I.A., Friedman, T.B., and Griffith, A.J. (2004). Genetic insights into the morphogenesis of inner ear hair cells. *Nature Reviews Genetics*, 489–498.

3. Lim, D.J. (1986). Functional structure of the organ of Corti: a review. *Hear Res 22*, 117–146.

4. Nayak, G.D., Ratnayaka, H.S., Goodyear, R.J., and Richardson, G.P. (2007). Development of the hair bundle and mechanotransduction. *Int J Dev Biol 51*, 597–608.

5. Denman-Johnson, K., and Forge, A. (1999). Establishment of hair bundle polarity and orientation in the developing vestibular system of the mouse. *J. Neurocytol*, 821–835.

6. Hudspeth, A.J., and Jacobs, R. (1979). Stereocilia mediate transduction in vertebrate hair cells (auditory system/cilium/vestibular system). *Proc Natl Acad Sci USA 76*, 1506–1509.

7. Cotanche, D.A., and Corwin, J.T. (1991). Stereociliary bundles reorient during hair cell development and regeneration in the chick cochlea. *Hear Res 52*, 379–402.

8. Tilney, M.S., Tilney, L.G., and DeRosier, D.J. (1987). The distribution of hair cell bundle lengths and orientations suggests an unexpected pattern of hair cell stimulation in the chick cochlea. *Hear Res 25*, 141–151.

9. Kaltenbach, J.A., and Falzarano, P.R. (1994). Postnatal development of the hamster cochlea. I. Growth of hair cells and the organ of Corti. *The Journal of comparative neurology 340*, 87–97.

10. Dabdoub, A., and Kelley, M.W. (2005). Planar cell polarity and a potential role for a Wnt morphogen gradient in stereociliary bundle orientation in the mammalian inner ear. *Journal of neurobiology 64*, 446–457.

11. Ross, A.J., May-Simera, H., Eichers, E.R., Kai, M., Hill, J., Jagger, D.J., Leitch, C.C., Chapple, J.P., Munro, P.M., Fisher, S., et al (2005). Disruption of Bardet-Biedl syndrome ciliary proteins perturbs planar cell polarity in vertebrates. *Nat Genet 37*, 1135–1140.

12. Jones, C., Roper, V.C., Foucher, I., Qian, D., Banizs, B., Petit, C., Yoder, B.K., and Chen, P. (2008). Ciliary proteins link basal body polarization to planar cell polarity regulation. *Nat Genet 40*, 69–77.

13. Kelley, M.W. (2006). Regulation of cell fate in the sensory epithelia of the inner ear. *Nat Rev Neurosci 7*, 837–849.

14. Dabdoub, A., Donohue, M.J., Brennan, A., Wolf, V., Montcouquiol, M., Sassoon, D.A., Hseih, J.C., Rubin, J.S., Salinas, P.C., and Kelley, M.W. (2003). Wnt signaling mediates reorientation of outer hair cell stereociliary bundles in the mammalian cochlea. *Development 130*, 2375–2384.

15. Ybot-Gonzalez, P., Savery, D., Gerrelli, D., Signore, M., Mitchell, C.E., Faux, C.H., Greene, N.D., and Copp, A.J. (2007). Convergent extension, planar-cell-polarity signalling and initiation of mouse neural tube closure. *Development 134*, 789–799.

16. Vladar, E.K., Antic, D., and Axelrod, J.D. (2009). Planar cell polarity signaling: the developing cell's compass. *Cold Spring Harb Perspect Biol 1*, a002964.

17. Wang, J., Mark, S., Zhang, X., Qian, D., Yoo, S.J., Radde-Gallwitz, K., Zhang, Y., Lin, X., Collazo, A., Wynshaw-Boris, A., et al (2005). Regulation of polarized extension and planar cell polarity in the cochlea by the vertebrate PCP pathway. *Nat Genet 37*, 980–985.

18. Montcouquiol, M., Rachel, R.A., Lanford, P.J., Copeland, N.G., Jenkins, N.A., and Kelley, M.W. (2003). Identification of Vangl2 and Scrb1 as planar polarity genes in mammals. *Nature 423*, 173–177.

19. Lu, X., Borchers, A.G., Jolicoeur, C., Rayburn, H., Baker, J.C., and Tessier-Lavigne, M. (2004). PTK7/CCK-4 is a novel regulator of planar cell polarity in vertebrates. *Nature 430*, 93–98.

Chapter 14

Role of Prickle1 and Prickle2 in Neurite Outgrowth in Murine Neuroblastoma Cells

Lisa Fujimura and Masahiko Hatano

Abstract

Murine Prickle2 but not Prickle1 gene expression was induced in C1300 neuroblastoma cell line during neurite-like process formation induced by all *trans*-retinoic acid (RA). Overexpression of Prickle1 or Prickle2 in C1300 cells induced striking neurite-like process formation without RA. Prickle1 and Prickle2 associate with Dishevelled1 (Dvl1) and overexpression of Prickle1 or Prickle2 resulted in the reduction of Dvl1 protein in C1300 cells. Overexpression of Dvl1 in C1300 cells prevented the neurite-like process formation induced by Prickle1 or Prickle2 overexpression. Prickle1 and Prickle2 promote neurite-like process formation of C1300 cells via the Dvl1-dependent mechanism.

Key words: Neuroblastoma cell line, Mouse cerebellum granule cell neurons, Prickle1, Prickle2, Dishevelled1, All *trans*-retinoic acid

1. Introduction

Murine Prickle1 and Prickle2 have been identified as homologues of Drosophila Prickle (1, 2). Prickle1 and Prickle2 have conserved three LIM domains, a PET domain, and a PKH domain. The LIM domains are found in many key regulators of developmental pathways and the PET domain is suggested to be involved in protein–protein interactions, whereas function of the PKH domain is not clear. Prickle1 and Prickle2 were mainly expressed in the neuronal cells during mouse embryogenesis (3, 4). Recent studies suggest that members of PCP family play important roles in neurite formation and axonal development (3, 5–7). Signaling pathways through Frizzled or Dvl coordinate cell and cytoskeletal polarity. Recently, Prickle1 and Prickle2 are also suggested to be involved in neurite formation in vitro. It was demonstrated that Prickle2 but not Prickle1 is strongly up-regulated during the RA-induced neurite-like process

Kursad Turksen (ed.), *Planar Cell Polarity: Methods and Protocols*, Methods in Molecular Biology, vol. 839,
DOI 10.1007/978-1-61779-510-7_14, © Springer Science+Business Media, LLC 2012

formation of C1300 murine neuroblastoma cells. Overexpression of Prickle1 or Prickle2 in C1300 cells induced spontaneous neurite-like process formation in the absence of RA. Furthermore, the effect of Prickle1 or Prickle2 on neurite-like process formation of C1300 cells was dependent on Dvl1 (8).

2. Materials

2.1. Cell Culture, Transfection, and Lysis

1. Dulbecco's modified Eagle's medium (DMEM) (Sigma) supplemented with 10% fetal bovine serum (FBS) (Sigma), 10 μg/ml streptomycin, 10 U/ml penicillin (SM/PC).

2. Solution of 0.25% trypsin (Sigma) and 1 mM ethylenediamine tetra acetic acid (EDTA) (Sigma).

3. RA (Sigma) is dissolved in ethanol at 1 mM, stored in aliquots at −80°C, and added to cell culture dishes at 1 μM as final concentration.

4. FuGENE 6 transfection kit (Roche Applied Science).

5. OPTI-MEM (GIBCO).

6. Lipofectamine RNAiMAX (Invitrogen).

7. NP-40 buffer for cell lysis: 50 mM Tris–HCl (pH 7.5), 100 mM NaCl, 1% NP-40, 5% Glycerol, 10 mM $MgCl_2$, 1 mM DTT, 1 mM Na_3VO_4, 10 mM NaF, 10 μg/ml Leupeptin.

8. Cell scrapers (IWAKI).

9. Tissue culture plate, 6-well (Falcon).

2.2. Cerebellar Granule Cells (CGNs) Culture

1. Digestion buffer: PBS with 0.1% trypsin, 0.2% EDTA.

2. Isolation media: DMEM supplemented with 10% BSA, SM/PC.

3. Isolation media containing of 100 μg/ml DNase1 (Sigma).

4. Culture media: Neurobasal-SFM Media (Sigma) supplemented with 2% B27 (GIBCO), 25 mM KCl, 2 mM glutamine, SM/PC.

5. Poly-D-lysine (PDL) (Sigma) is diluted in PBS at 0.001%, and coat culture dishes. After coat the dishes with PDL solution for 10 min, rinse twice with DDW and air dry for 1 h.

6. Tissue culture plate, 6-well (Falcon).

2.3. RNA Isolation, cDNA Synthesis, and Real Time (RT)-PCR

1. TRIzol Reagent (Invitrogen).

2. Super-Script first-strand synthesis system for RT-PCR (Invitrogen).

3. RNaseOUT Recombinant RNase Inhibitor (Invitrogen).

4. SYBR Green PCR Master Mix (PE Applied Biosystems).

5. ABI Prism 7000 Sequence Detection System (PE Applied Biosystems).

**2.4. Immuno-
histochemistry**

1. 4% Paraformaldehyde (PFA).

2. Permeabilization solution: 0.3% Triton-X-100 in PBS.

3. Blocking buffer: 5% BSA in PBS.

4. Antibody dilution buffer: 1% BSA in PBS.

5. First antibody: monoclonal anti-tubulin antibody (Clone 6-11B-1, 1:1,000; Sigma).

6. Second antibody: Cy3-conjugated anti-mouse IgG antibody (1:2,000; Jackson ImmunoResearch).

7. Flourmount (Diagnostic Biosystems).

**2.5. Immuno-
precipitation**

1. NP-40 buffer for cell lysis: 50 mM Tris–HCl (pH 7.5), 100 mM NaCl, 1% NP-40, 5% Glycerol, 10 mM $MgCl_2$, 1 mM DTT, 1 mM Na_3VO_4, 10 mM NaF, 10 µg/ml Leupeptin.

2. Protein G-Sepharose 4FF (Amersham Bioscience).

3. Mouse anti-Dvl1 antibody (Santa Cruz Biotechnology).

4. Five times SDS sample buffer: 156 mM Tris–HCl (pH 7.5), 50% Glycerol, 25% 2-melcaptoethanol, 10% SDS, several drop of Bromophenol Blue.

**2.6. SDS–
Polyacrylamide Gel
Electrophoresis**

1. 6 ml Separating gel: 15% bis/acril amide, four times gel buffer (1.5 M Tris–HCl pH 8.8, 0.4% SDS), 72 µl 10% Ammonium persulfate (APS), 12 µl TEMED.

2. 3 ml Stacking gel: 4% bis/acril amide, four times gel buffer (0.5 M Tris–HCl pH 6.8, 0.4% SDS), 30 µl 10% APS, 5 µl TEMED.

3. Running buffer: 0.1% SDS, 0.4 M Glycine, 0.6% Tris–HCl.

4. Prestained SDS–polyacrylamide gel electrophoresis (SDS–PAGE) standards, Low Range (BIO-RAD).

2.7. Western Blotting

1. Transfer buffer: 5 mM Tris-HCl, 20 mM Grycine, 0.1% SDS, 20% ethanol.

2. Immobilon-P filters (Millipore).

3. 1% Blocking solution: 1% Block Ace powder (DS PHRMA BIOMEDICAL) in TBS (20 mM Tris–HCl, 136 mM NaCl, pH 7.6).

4. Primary antibody dilution buffer: 0.1% blocking solution in TBS.

5. Washing buffer: 0.1% Tween 20 in TBS.

6. First antibody: mouse anti-FLAG M2 monoclonal antibody (1:1,000) (Sigma), mouse anti-Dvl1 antibody (1:1,000) or anti-actin goat polyclonal IgG antibody (1:500) (Santa-Cruz Biotechnology).

7. Second antibody: horseradish peroxidase (HRP)-conjugated anti-mouse IgG (1:10,000) (Jackson Immunoresearch) or HRP-conjugated anti-Goat Immunoglobulins (1:5,000) (DAKO).

8. Western Lightning plus ECL (PerkinElmer Inc).

3. Methods

It is known that RA stimulation of C1300 mouse neuroblastoma cell line induces neurite-like process extension. Expression of Prickle2 but not that of Prickle1 was induced from day 1 and gradually increased until day 3 after stimulation. On the other hand, primary culture of mouse cerebellum granule cell neurons (CGNs) rapidly starts to grow and produce processes. Prickle1 but not Prickle2 expression was induced during morphological differentiation in vitro. Prickle1 and Prickle 2 are differentially expressed in neurons during morphological differentiation. Thus, it is important to know which PCP genes are expressed in each experimental system.

Overexpression of Prickle1 or Prickle2 leads to the induction of striking morphological changes such as neurite-like process formation in C1300 cells. Transfection of siRNA for Prickle2 in C1300 cells resulted in a reduction of neurite bearing cells compare with control siRNA.

Expression of Dvl1 mRNAs was strongly induced in C1300 cells 2 days after RA stimulation, but the amount of protein was reduced. Overexpression of Prickle1 or Prickle2 in C1300 cells reduced the amount of Dvl1 in a dose-dependent manner (see Fig. 3). We showed that murine Prickle 1 and Prickle2 can interact with Dvl1 (see Fig. 4) and may function through its degradation. Indeed, Dvl1 overexpression in C1300 cells significantly reduced the percentage of neurite-like process bearing cells among transfectants. When Dvl1 was co-expressed with Prickle1 or Prickle2, neurite-like process formation induced by Prickle overexpression was suppressed.

In this chapter, we describe the standard protocol to detect PCP gene mRNA expression in cultured cells. Next, we show morphological analysis and neurite extension assay using neuroblastoma cell line. Finally, analysis of protein–protein interaction using immunoprecipitation assay is described.

3.1. Analysis of PCP Gene Expression by Real Time (RT)-PCR

3.1.1. Isolation of Total RNA

1. Pellet cells by centrifugation ($270 \times g$) at 4°C.
2. Lyse the cells in 1 ml of TRIzol Reagent and incubate for 5 min at room temperature (see Note 1).
3. Add 0.2 ml of chloroform and shake it vigorously for 15 s. Incubate for 3 min at room temperature.
4. Centrifuge ($20,000 \times g$) for 15 min at 4°C.
5. Transfer the upper aqueous phase to a 1.5-ml Eppendorf tube.
6. Add 0.5 ml of isopropyl alcohol. Mix and incubate for 10 min at room temperature.
7. Centrifuge ($20,000 \times g$) for 10 min at 4°C.
8. Remove the supernatant.

9. Wash the RNA pellet with 75% ethanol.

10. Centrifuge $(20,000 \times g)$ for 5 min at 4°C.

11. Remove the supernatant.

12. Resolve the RNA pellet with 10–20 μl of DEPC-DDW and incubate for 10 min at 55°C (see Note 2 and 3).

3.1.2. cDNA Synthesis   1. Add the following components to 0.2 ml tube.

500 μg/ml oligo (dT)	1 μl
Total RNA	1–5 μg
10 mM dNTP mix	1 μl
DEPC-DDW (Adjust total volume to 13 micro-litter with DEPC DDW)	
Total	13 μl

2. Heat mixture at 65°C for 5 min and incubate on ice for 2 min.

3. Add the following components to the mixture 1.

5×First-strand buffer	4 μl
0.1 M DTT	1 μl
RNaseOUT	1 μl
SuperScriptIII	1 μl
Total	20 μl

4. Incubate at 50°C for 60 min.

5. Inactivate the reaction by incubating at 70°C for 15 min.

3.1.3. Real Time (RT)-PCR   Primers are as Table 1.

1. Put the following mixture into 96-well plates.

2×SYBR Green PCR Master Mix	12.5 μl
10 μM primer (forward)	0.75 μl
10 μM primer (reverse)	0.75 μl
DDW	1 μl
Total	15 μl

2. Add 10 μl of each cDNA. Dilute each cDNA as follows.

Sample: dilute cDNA one fifth with DDW.

Standard: dilute β-actin cDNA one fifth with DDW and make dilution cascade (dilute one half for six times).

Table 1
RT-PCR primers

Prickle1	Forward	5′-TGAGAATGTCCACGCGATGAG-3′
	Reverse	5′-TAAAGGCCAACAGCAAGTTGGA-3′
Prickle2	Forward	5′-AGCATCCCAGTTCCCAAGTATGAG-3′
	Reverse	5′-GGCACCACCTTCAGCAGACA-3′
Dvl1	Forward	5′-ACTTCAAGAACGTGCTCAGCAA-3′
	Reverse	5′-TGGCATTGTCATCGAAGATCTC-3′
Dvl2	Forward	5′-AGTGTCACCGATTCCACAATGTC-3′
	Reverse	5′-TCGTTACTTTGGCCCACAATG-3′
Dvl3	Forward	5′-AAAGCAATGAGCGAGGTGATG-3′
	Reverse	5′-ATCGTTGCTCATGTTCTCGAAG-3′
β-Actin	Forward	5′-CCAGCCTTCCTTCTTGGGTAT-3′
	Reverse	5′-TGGCATAGAGGTCTTGGCATAGAGGTCTTTACGGATGT-3′

3. Run on an ABI Prism 7000 Sequence Detection System. Cycling steps are as follows:

   Step 1: 95°C for 15 s.

   Step 2: 60°C for 1 min.

   Repeat steps 1 and 2 for 40 times.

4. Analyze the data using Sequence Detector software (PE Applied Biosystems). The data are reported as the ratio of the calculated amount of candidate RNA in a given sample by the calculated amount of the housekeeping control gene (β-actin) in the same sample. Example of the expression is shown in Fig. 1.

3.2. Neurite Extension in C1300 Cells and CGNs

3.2.1. C1300 Cells

1. C1300 cells are maintained in DMEM (10% FBS) on 100-mm tissue culture dishes and are passaged with Trypsin/EDTA.

2. Seed the 2 ml of 0.5×10^5 cells/ml into 6-well tissue culture plates.

3. Change the medium to 2 ml of DMEM (3% FBS) next day.

4. Add 1 μM of RA to induce neurite extension.

5. Add the RA on everyday till 72 h.

3.2.2. CGNs

1. Mince the cerebella dissected from P7 mice into 500 μm into 2 ml of PBS in 35-mm dish (see Note 4).

2. Transfer the 2 ml of minced tissue into 15-ml tube and add 4 ml of digestion buffer. Incubate for 10 min at 37°C. (Warm the digestion buffer to 37°C before addition.)

3. Add the 8 ml of isolation media and centrifuge ($270 \times g$) at room temperature for 5 min.

4. Suck the media.

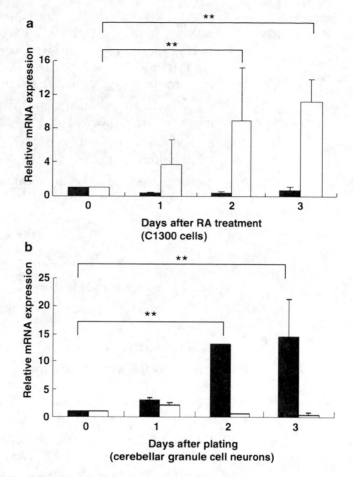

Fig. 1. Expression of Prickle1 or Prickle2 during neurite-like process formation. (a) Prickle2 but not Prickle1 expression was induced in C1300 cells during neurite-like process formation by RA. (b) Prickle1 but not Prickle2 expression was induced in cerebellar granule cell neurons cultured in the presence of KCl. Both in (a) and (b), the data are expressed as relative to the level of day 0. Closed bars; Prickle1, open bars; Prickle2. Values are means ± SD of three independent experiments (** $P < 0.01$).

5. Resuspend in 2 ml of isolation media containing DNase1.

6. Triturate the suspension through a fire-polished glass pipette for ten times on ice.

7. Add the 10 ml of isolation media and centrifuge ($270 \times g$) at room temperature for 5 min.

8. Suck the media.

9. Resuspend the tissue pieces in culture media at 3×10^5 cells/ml.

10. Seed the 2 ml of cells onto 6-well tissue culture plates coated with PDL.

11. Culture plates at 37°C in a 5% CO_2 atmosphere.

12. CGNs rapidly start to grow and produce processes. After 2–3 days of in vitro culture, the meshwork of fine processes appears.

3.3. Morphological Analysis of C1300 Cells

3.3.1. Transfection for Overexpression

1. Prickle1-GFP or Prickle2-GFP (pCMV-Pk1-IRES-GFP, pCMVPk2-IRES-GFP), DsRed-tagged Dvl1 (pCMV-DsRed-Dvl1) expression plasmids were constructed by PCR amplifying Prickle1, Prickle2, or Dvl1 cDNA fragments and ligating them to the multiple cloning site of each vector.

2. C1300 cells are maintained in DMEM (10% FBS) on 100-mm tissue culture dishes and are passaged with Trypsin/EDTA.

3. Seed the 1×10^5 cells/well in DMEM (5% FBS) into 6-well tissue culture plates.

4. Next day, add the following components to 1.5-ml Eppendorf tube (for 1 well).

OPTI-MEM	100 μl
FuGENE6	6 μl

Mix by pipetting and incubate for 5 min at room temperature.

5. Add 1 μg of plasmid and mix, incubate for 15 min at room temperature.

6. Add the mixture into the well.

7. Culture plates at 37°C in a 5% CO_2 atmosphere for 48 h.

3.3.2. Transfection for Knockdown

1. The nucleotide sequences of Small interference RNA (siRNA) against mouse Prickle2 and control are as Table 2.

2. C1300 cells are maintained in DMEM (10% FBS) on 100-mm tissue culture dishes and are passaged with Trypsin/EDTA.

3. Add the following components to 1.5-ml Eppendorf tube (for one well).

OPTI-MEM	500 μl
siRNA	30 pmol
Lipofectamine RNAiMAX	5 μl

Mix by pipetting and incubate for 15 min at room temperature.

Table 2
Small interference RNA (siRNA) against mouse Prickle2 and control

Prickle2 siRNA	5'-GCAAGCUCAUGUUUGACUUTT-3'
Negative control (scrambled siRNA for Prickle2)	5'-UAUGGUGUACCAACCUGUUTT-3'

4. Seed the 1×10^5 cells/2.5 ml in DMEM [10% FBS, SM/ PC(–)] into 6-well tissue culture plates.

5. Add the mixture into the well.

6. Culture plates at 37°C in a 5% CO_2 atmosphere for 48 h (see Note 5).

3.3.3. Immuno-histochemistry

1. Plate a fire-polished microcover glass (Matsunami) into each of the 6-well tissue culture plates (all procedures are done using the same plates).

2. Seed the C1300 cells onto a microcover glass and transfect with various expression vectors.

3. After 48 h of culture, suck the media and fix the cells with 1 ml of 4% PFA for 15 min (see Note 6).

4. Suck the 4% PFA and wash with 2 ml of PBS 5 min for three times.

5. Suck the PBS and permeabilize the cells with 1 ml of permeabilization solution for 5 min.

6. Suck the solution and wash with PBS 5 min for three times.

7. Suck the PBS and block with blocking buffer for 1 h at room temperature.

8. Suck the blocking buffer and expose to the 50–100 μl of first antibody overnight at 4°C (see Note 7).

9. Wash with PBS 5 min for three times.

10. Suck the PBS and expose to 50–100 μl of second antibody for 1 h at room temperature.

11. Wash with PBS 5 min for three times.

12. Suck the PBS and put aside the microcover glass carefully and Mount the cells on slide grass (Matsunami) with Flourmount (see Note 8).

13. The cells are observed under a fluorescence microscope (Axioplan 2 imaging, Carl Zeiss) and photographed (AxioCam, Carl Zeiss). Morphological analysis is done using NIH image software (ImageJ 1.37v). Measurements used to estimate neurite-like process formation are percentage of neurite-like process bearing cells and average length of neurite-like process with neurite defined as a process longer than the cell body. More than 100 cells are evaluated for each experiment. Examples of the immunohistochemistry are shown in Fig. 2.

3.4. Interaction of Prickle1 and 2 with Dvl1

3.4.1. Immunoprecipitation

1. HEK293 cells (1×10^7) transiently transfected with plasmids FLAG-tagged Prickle1 or Prickle2 (pCR3.1-2FLAG-Pk1, pCR3.1-2FLAG-Pk2), HA-tagged Dvl (pCR3.1-2HA-Dvl1) are lysed at 4°C in 0.8 ml of NP-40 buffer.

2. Clarify the lysates by centrifugation ($270 \times g$) at 4°C.

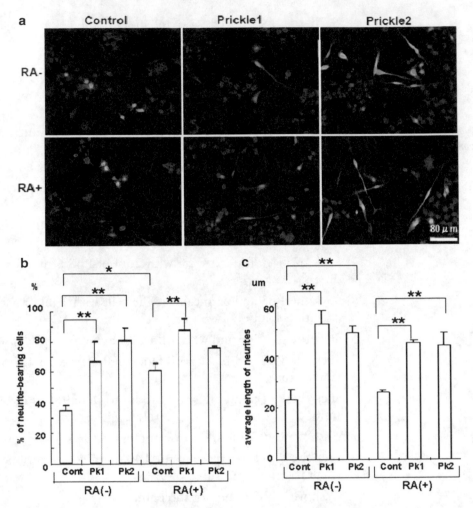

Fig. 2. Overexpression of Prickle1 or Prickle2 promotes neurite-like process formation of C1300 cells. (**a**) Morphologies of Prickle1 (Pk1) or Prickle2 (Pk2) transfected cells. C1300 cells were transfected with Prickle1 or Prickle2 expression plasmids and cultured with (RA+) or without (RA−) RA for the last 24 h. The cells were fixed and immunostained 48 h after transfection. Microtubule was visualized by immunostaining with anti-tubulin antibody (*red*). Control: GFP-vector transfected cells. (**b**) Numbers of neurite-like process bearing cells shown in panel (**a**) were quantified. Neurite-like process bearing cells among effectively transfected (GFP positive) cells were calculated. (**c**) The length of neurite-like process shown in panel (**a**) was measured and quantified. Average neurite-like process length among effectively transfected cells was calculated. Values are means ± SD of three independent experiments (*$P < 0.05$, **$P < 0.01$). (For interpretation of the references to color in this figure legend, the reader is referred to the web version of the article (8)).

3. Rotate the supernatant at 4°C with 5 μl of protein G-Sepharose 4FF for 30 min (pre-clear) (see Note 9).

4. After centrifugation ($270 \times g$) at 4°C, rotate the supernatant with antibody at 4°C for 1 h.

5. Add 10 μl of protein G-Sepharose 4FF and rotate at 4°C for 1 h.

6. Centrifuge ($270 \times g$) the samples at 4°C for 1 min and wash the pellets four times with 1 ml of lysis buffer.

7. Resolve in 20 μl of three times of SDS sample buffer, boil for 5 min, and fractionate by 15% SDS–PAGE. Example is shown in Fig. 4.

3.4.2. SDS–PAGE

1. Prepare 1.5-mm thick glass plates for the gels (NA1021) (Nihon EIDO, Japan).

2. Prepare 6 ml of the separating gel and pour into the space. Overlay with DDW and leave it for 30 min.

3. Pour off the DDW and rinse the top of the gel with water.

4. Prepare 3 ml of the stacking gel and pour into the space. Quickly insert the comb. Leave it for 30 min.

5. Carefully leave the comb and set it to the chamber for electrophoresis.

6. Add the running buffer to the upper and lower chambers.

7. Wash the wells with running buffer.

8. Load the sample and marker into the wells.

9. Run the gels at 300 V, 30 mA for 1 h (my Power300) (ATTO, Japan).

3.4.3. Western Blotting

1. SDS–PAGE followed by electroblotting onto filters with transfer buffer 50 V for 90 min (Marisol, Cat #KS-8451).

2. Block the filters for 1 h with 1% blocking solution.

3. Incubate the filters into hybridization bag with a first antibody overnight at 4°C.

4. Wash the filters with washing buffer 5 min for three times.

5. Incubate the filters with a second antibody for 1 h at room temperature.

6. Wash the filters with washing buffer 5 min for three times.

7. Visualize the Immunoreactive bands using ECL detection reagents (Amersham Biosciences). Chemiluminescent signal can be detected digitally wth an instrument such as Light-capture II™ (ATTO, Japan). Examples are shown in Figs. 3 and 4.

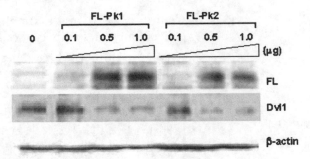

Fig. 3. Prickle1 and Prickle2 interact with Dvl1. Overexpression of Prickle1. (FL-Pk1) or Prickle2 (FL-Pk2) reduced the endogenous Dvl1 protein in C1300 cells. Prickle1 or Prickle2 expression plasmid was transfected into C1300 cells at the indicated amounts and proteins were isolated 48 h after transfection. Western blot analysis was performed using anti-FLAG antibody (FL) and anti-Dvl1 antibody (Dvl1).

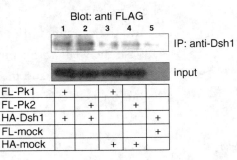

Fig. 4. Prickle1 and Prickle2 interact with Dvl1. Coimmunoprecipitation of Prickle1 or Prickle2 with Dvl1. HA-tagged-Dvl1 (HA-Dvl) was co-transfected with Prickle1 (FL-Pk1) or Prickle2 (FL-Pk2) into HEK293 cells. Dvl1 in cell lysates was immunoprecipitated with anti-Dvl1 antibody and immunoblotted with anti-FLAG antibody. Empty vectors (FL-mock and HA-mock) were used as a negative control. The cell lysate was used as an input control.

4. Notes

1. After lysis with TRIzol reagent, samples can be stored at –60 to –70°C for at least 1 month.

2. If the pellet of RNA precipitate is very small, add the 10 µl of DEPC-DDW and use all samples for cDNA synthesis.

3. The RNA can be stored at –80°C for 1 year.

4. To avoid bacterial contamination, dry-heat sterilize the scissors and tweezers.

5. To assess and optimize the efficiency of siRNA into C1300 cells, use the BLOCK-iT Alexa Flour Red Fluorescent Oligo (Invitrogen). It allows easy fluorescence indication.

6. To avoid the cells coming off from the bottom of the plate, pour the liquid with pippet gently.

7. To avoid the samples drying up, use the moisture chamber and pour the water into it.

8. The samples can be stored at 4°C in dark several months.

9. Before using the protein G-Sepharose 4FF, it has to be neutralize with NP-40 buffer.

Acknowledgments

The authors would like to thank A. Goda and M. Hanazono for skillful technical assistance and H. Takano for discussion. This work was supported in part by Grants-in-Aid from the Ministry of Education, Science, Technology, Sports, and Culture of Japan.

References

1. D. Gubb, C. Green, D. Huen, D. Coulson, G. Johnson, D. Tree, S. Collier, J. Roote (1999) The balance between isoforms of the Prickle LIM domain protein is critical for planar polarity in Drosophila imaginal discs, *Genes Dev.* **13**, 2315–2327.

2. M. Katho, M. Katho (2003) Identification and characterization of human PRICKLE1 and PRICKLE2 genes as well as mouse Prickle1 and Prickle2 genes homologous to Drosophila tissue polarity gene prickle, *Int. J. Mol. Med.* **11**, 249–256.

3. H. Okuda, S. Miyata, Y. Moria, M. Tohyama (2007) Mouse Prickle1 and Prickle2 are expressed in postmitotic neurons and promote neurite outgrowth, *FEBS Lett.* **581**, 4754–4760.

4. F. Tissir, A.M. Goffinet (2006) Expression of planar cell polarity genes during development of the mouse CNS, *Eur. J. Neurosci.* **23**, 597–607.

5. L.V. Goodrich (2008) The plane fact of PCP in the CNS, *Neuron* **60**, 9–16.

6. S. Kishida, H. Yamamoto, A. Kikuchi (2004) Wnt-3a and Dvl induce neurite retraction by activating Rho-associated kinase, *Mol. Cell Biol.* **24**, 4487–4501.

7. F. Tissir, I. Bar, Y. Jossin, O. De Backer, A.M. Goffinet (2005) Protocadherin Celsr3 is crucial in axonal tract development, *Nat. Neurosci.* **8**, 451–457.

8. L. Fujimura, H. W-Takano, Y. Sato, T. Tokuhisa, M. Hatano (2009) Prickle promotes neurite outgrowth via the Dishevelled dependent pathway in C1300 cells, *Neurosci. Lett.* **467**, 6–10.

Chapter 15

The Planar Cell Polarity Pathway and Parietal Endoderm Cell Migration

Kristi LaMonica and Laura Grabel

Abstract

Parietal endoderm (PE) migration is the first long-range migratory event in the mammalian embryo contributing to the parietal yolk sac. PE migration can be studied in vitro using the F9 teratocarcinoma stem cell model system. We have found that PE migration is directed and modulated via the Planar Cell Polarity (PCP) pathway through Rho/ROCK signaling. Wnt inhibition using sFRP results in a loss of orientation, visualized by Golgi apparatus localization, along with disorganized microtubules and a lack of robust focal adhesions. Small GTPases are downstream of PCP signaling and Rho/ROCK inhibition results in a loss of orientation, whereas inhibition of Rac does not affect PCP. Activation of canonical Wnt signaling combined with Wnt inhibition does not prevent loss of oriented migration. These data support a role for non-canonical Wnt/PCP signaling directing oriented migration of PE.

Key words: Planar cell polarity, Parietal endoderm, Oriented migration, Polarized migration

1. Introduction

Although the planar cell polarity (PCP) pathway was first identified based on its role in polarizing epithelial structures such as bristles and hairs, the components of this cascade are also involved in promoting cell migration (1–3). Data suggest a role in directing motility of individual cells, for example during the epithelial-to-mesenchymal transition that characterizes neural crest migration, as well as in collective migration of cell sheets for example during convergent extension movements in gastrulating embryos (1, 3–5). We have established a role for the PCP pathway, acting via the small GTPase Rho, in parietal endoderm (PE) migration using an F9 teratocarcinoma or embryonic stem cell-based embryoid body outgrowth system.

Kursad Turksen (ed.), *Planar Cell Polarity: Methods and Protocols*, Methods in Molecular Biology, vol. 839,
DOI 10.1007/978-1-61779-510-7_15, © Springer Science+Business Media, LLC 2012

In the embryo, this migration event is essential for proper yolk sac formation, which supplies the embryo with nutrients and oxygen early in gestation, prior to formation of the placenta (6). In the in vitro model system, the outer layer visceral endoderm cells migrate as a sheet away from the attached embryoid body and onto a fibronectin-coated substrate (7). The cells can migrate in an oriented manner, with Golgi situated in front of the nucleus in the direction of migration (Fig. 1) or, if the PCP pathway is inhibited, in a disoriented fashion (Fig. 1). Outgrowth under PCP-inhibited conditions is more extensive, likely due to the absence of strong adhesion to the extracellular matrix.

Based on a series of pharmacological or genetic interventions, described in detail below, we propose that ligand binding to the Frizzled (Fz) receptor recruits Disheveled (Dsh), which acts via Daam1 and Rho to activate Rho-dependent kinase (ROCK). ROCK in turn phosphorylates a number of substrates, including Myosin Light Chain Phosphatase (MLCP) and Myosin Light Chain (MLC) itself, resulting in MLC activation, required for engagement of the cytoskeleton and oriented migration (Fig. 2). We also describe approaches to test potential contributions of the Rac-dependent arm of the PCP and the canonical β-catenin-dependent Wnt pathway to PE migration.

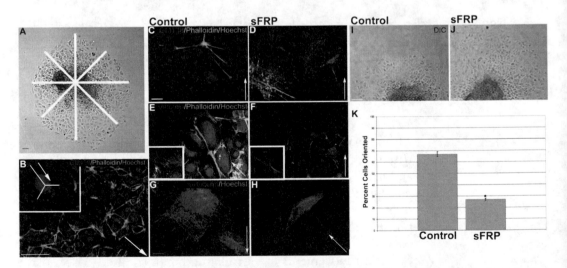

Fig. 1. Oriented migration. (a) 4× DIC image of an embryoid body and surrounding outgrowth. The *white lines* superimposed on the image show the lines used to measure migration extent. (b) Representation of an oriented control outgrowth. *White arrow* denotes direction of migration. The *inset* demonstrates how to measure orientation, the cell is divided into thirds as shown and cells are considered orientation if over 50% of the Golgi apparatus (*red*) is localized in front of the nucleus in the direction of migration. (c) Control outgrowth exhibits oriented migration with Golgi apparatus localized in front of the nucleus in the direction of migration, whereas sFRP-treated outgrowths display a lack of oriented migration (d, k). sFRP-treated outgrowth also has reduced focal adhesions (f) relative to controls (e), which leads to increased migration extent as also observed in DIC (i, j). We also observe microtubule alignment in the direction of migration in control outgrowths (g), whereas microtubules are disorganized and often perpendicular to the direction of migration in sFRP-treated outgrowths (h). *White arrows* denote direction of migration. Reproduced from LaMonica et al., 2009 with permission from Elsevier.

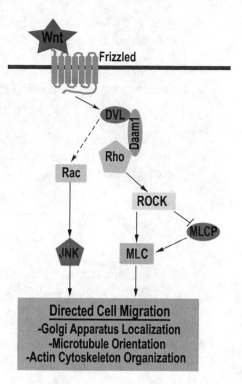

Fig. 2. PCP pathway. The Wnt ligand binds to the Frizzled receptor, which then activates DVL. DVL interacts with Daam1, activating the small GTPase Rho. Rho modulates downstream activity via ROCK indirectly through MLCP, or directly through MLC to promote oriented migration. Alternatively, DVL can activate the small GTPase Rac, which can modulate oriented migration though JNK.

Wnt signaling can be inhibited at the level of Wnt using the soluble Frizzled receptor protein (sFRP), which is added to the tissue culture medium. sFRP acts by binding secreted Wnts thereby preventing Wnt from binding to the cell anchored Frizzled protein. sFRP inhibition of Wnt signaling is not pathway specific, or selective for canonical versus non-canonical. To assay for canonical versus non-canonical signaling, it is necessary to inhibit or activate the various arms of the pathways at different levels. The Rho arm of the pathway can be inhibited at the level of ROCK using the cell-permeable pharmacological inhibitor Y27632, while the Rac arm of the pathway can be inhibited at the level of c-jun kinase (JNK) using the pharmacological inhibitor SP600125 (8–15). Activity of both inhibitors can be assayed via western blotting for MLCP and phospho-JNK, respectively. To provide the link between PCP signaling and the Rho pathway, Daam1 activity can be genetically modulated. The amino terminus (N-Daam) inhibits, whereas the carboxy portion of the protein (C-Daam) acts as to constitutively activate downstream signaling (16, 17). To provide a link between PCP signaling through the Rho arm of the pathway, cells can be transiently transfected with C-Daam and treated 24 h post-transfection with sFRP (14). If PCP signaling is modulated through the Rho arm of the pathway, C-Daam transfection will protect the cells from perturbation of PCP from sFRP treatment. To test the contributions of the canonical Wnt pathway, we inhibited GSK-3β using the BIO inhibitor, which results in an increase of active, non-phosphorylated β-catenin (18). Levels of active β-catenin and total β-catenin can be assayed by western blotting under treated and control conditions. If the canonical pathway is involved in modulating PCP, increased β-catenin should protect cells from sFRP-mediated loss of oriented migration (14).

2. Materials

2.1. Cell Culture

1. Dulbecco's Modified Eagle's Medium (DMEM) (Sigma) supplemented with 10% fetal bovine serum (FBS) (Hyclone, Ogden, Utah), 1% L-glutamine (Invitrogen), and 1% penicillin/ streptomycin (Invitrogen).
2. Tryple Express (Invitrogen) (see Note 1).
3. Porcine gelatin, cell culture grade (Sigma).
4. 7.5×10^{-7} M *Trans*-retinoic Acid (Sigma).
5. 30 μg/ml of 1 mg/ml fibronectin (Sigma).

2.2. Immunocyto-chemistry

1. Phosphate Buffered Saline (PBS), 3.7% Formaldehyde (from 37% stock) (Sigma), and 1% Bovine Serum Albumin in PBS (BSA, Sigma).
2. Primary antibodies: mouse monoclonal anti-GM130 (BD Biosciences), mouse monoclonal anti-α-tubulin (Sigma), and mouse monoclonal anti-vinculin (Sigma).
3. Phalloidin (Invitrogen): Rhodamine phalloidin, Alexa 488 phalloidin, and Alexa 647 phalloidin.
4. Alexa-fluor secondary antibodies (Invitrogen): goat-anti-mouse Alexa 568 and goat-anti-mouse Alexa 488.
5. Hoechst 33342 (Invitrogen).
6. Gelvatol/NPG or other anti-fade/anti-photobleaching mounting medium.

2.3. Outgrowth Assay, Extent, and Orientation

1. Hoechst 33342-stained outgrowth.
2. Microscope, camera, Adobe Photoshop.
3. GM130, phalloidin, and Hoechst 33342-stained outgrowth.

2.4. PCP Inhibitors

1. 25 mg/ml sFRP (R&D Systems) reconstituted in 1 ml of sterile water used at 2.5 μg/ml.
2. 1 mg/ml Y27632 (EMD Biosciences/Calbiochem). Reconstitute in 1 ml sterile water.

3. 50 mM in solution SP600125 (EMD Biosciences/Calbiochem). Make a 1:100 dilution of an aliquot for a 0.5 mM working solution.

4. 1 mg/ml GSK-3β inhibitor 6-bromoindirubin-3′-oxime (BIO) (EMD Biosciences/Calbiochem) diluted in 2.85 ml 5% DMSO and sterile water to a 1 mM solution.

2.5. Transfection

1. Lipofectamine 2000 (Invitrogen).

2. Antibiotic-free F9 Medium (2.1.1 minus penicillin/streptomycin).

3. Fibronectin-coated coverslips or gelatin-coated dishes.

4. Midiprepped plasmids for transient transfection. C- and N-Daam constructs were generous gifts from Habas and coworkers (16, 17). The C3 transferase construct to inhibit Rho signaling was a generous gift from Burridge and coworkers (19, 20).

2.6. Western Blotting

1. Lysis buffer containing 50 mM Tris–HCl pH 7.4, 150 mM NaCl, 1 mM $CaCl_2$ 1% NP-40, 0.1% SDS, 0.5% sodium deoxycholate and add protease and phosphatase fresh each time (10 μg/ml leupeptin, pepstatin, aprotinin, and benzamidine HCL; 1 mM PMSF and sodium orthovanadate).

2. Bradford Reagent (BioRad) or Precision Plus Red (Cystoskeleton) for protein quantification.

3. Running Buffer (10× Laemmli) for 1 L stock (30.2 g Tris Base, 144.1 g Glycine, 10 g SDS). Dilute to 1× with distilled water. 1 L is needed for both Invitrogen and BioRad mini gel systems.

4. 2× Loading buffer containing 20 mM Tris–HCL pH 6.8, 20% SDS, 2 g sucrose, 200 μl β-mercaptoethanol, 5 mg bromophenol blue, 10 mg methylene blue chloride. Aliquot and store at –20°C. A commercially available loading buffer supplemented with β-mercaptoethanol can also be used.

5. Transfer buffer (1 L) composed of 6.007 g Tris Base, 28.82 g glycine, 1 ml 10% SDS, and 100 ml methanol.

6. TBS-Tween (1 L) wash buffer containing 150 ml 1 M NaCl, 10 ml Tris–HCL pH 7.4, 3 ml Tween-20.

7. Blocking buffer composed of 3% non-fat powdered milk (grocery store brand) and 0.5% bovine serum albumin (Sigma) in TBS-tween.

8. Stripping Buffer made of 62.5 mM Tris–HCl pH 6.7, 2% SDS, 100 mM β-mercaptoethanol. Add β-mercaptoethanol prior to use to volume needed for stripping.

9. Precision Plus Blue prestained molecular weight standard (BioRad).

10. Mini gel running and transfer apparatus (BioRad or Invitrogen), Immobilon P (Millipore), Kodak BioMax Light x-ray film, x-ray cassette, and x-ray film developer.

11. Primary antibodies: mouse anti-active β-catenin (1:1,000) (Millipore), rabbit anti-β-catenin (1:5,000) (Cell Signaling), mouse anti-α-fodrin/spectrin (1:1,000) (MP Biomedicals), rabbit anti-myosin light chain phosphatase (1:1,000) (Upstate), rabbit anti-phospho-myosin light chain phosphatase (1:1,000) (Covance), rabbit anti-c-jun kinase (1:500) (Santa Cruz, and mouse anti-phospho c-jun kinase (1:200) (Santa Cruz).

12. HRP-Conjugated Secondary antibodies: goat anti-mouse HRP and goat anti-rabbit HRP diluted 1:10,000 (Sigma).

3. Methods

3.1. Cell Culture

3.1.1. Passaging F9 Stem Cells

1. Coat appropriate number of 6 cm tissue cultures dishes with 2 ml of gelatin per dish making sure that the bottom of the dish is fully covered.

2. Aspirate old media off the dishes.

3. Wash the cells with 2 ml of PBS per dish.

4. Aspirate off PBS.

5. Add 2 ml Tryple Express to each dish and incubate at 37°C for 5 min.

6. Quench the Tryple Express with 2 ml of media per dish.

7. Pipette the cells with the Tryple Express and media into a 15-ml conical tube.

8. Spin in a clinical centrifuge at 3,000 rpm for 3 min. (There should be a very visible pellet of stem cells at the bottom of the conical tube.)

9. Aspirate off media and Tryple Express.

10. Resuspend the cell pellet in 5 ml of media.

11. Aspirate the gelatin off the tissue culture dishes.

12. Add 5 ml of media per tissue culture dish and add 250 µl of the resuspended cell pellet to each dish. F9 Stem cells usually need to be passaged every fourth day if the day they were last passed is considered day 1. Change media on day 3 to fresh media.

3.1.2. Making Embryoid Bodies

1. Add 10 ml of media to the desired number of 10 cm petri dishes.

2. Add 600 µl of resuspended stem cells from passaging.

3. Add 10 µl retinoic acid to each dish for a final concentration of 7.5×10^{-7} M.

4. Continue to add retinoic acid to each dish of embryoid bodies daily. Label new dishes of embryoid bodies as day 1.

3.1.3. Feeding Embryoid Bodies

1. Pipette embryoid bodies from the Petri Dish into a 15-ml conical tube.

2. Allow the embryoid bodies to settle.

3. Aspirate off the old media. Take care not to disturb the pelleted embryoid bodies.

4. Resuspend the embryoid bodies in 10 ml of media.

5. Put resuspended embryoid bodies into a new 10-cm Petri dish.

6. Add 10 µl retinoic acid for a final concentration of 7.5×10^{-7} M.

7. Add 10 µl retinoic acid daily. Embryoid bodies usually need their media changed on day 4 and day 6 if the day of passaging is considered day 1.

3.1.4. Plating Embryoid Bodies

1. Wash coverslip in a petri dish with 0.5 ml phosphate buffered saline (PBS) for 5 min at room temperature (RT). The ECM will spread over the coverslip better if the PBS does not spill over onto the Petri dish from the coverslip.

2. Aspirate PBS.

3. Coat coverslips with 100 µl of 30 µg/ml fibronectin diluted in PBS for 4 h or overnight at 37°C in a tissue culture incubator.

4. Aspirate excess fibronectin and plate approximately 50 embryoid bodies on the coverslip (or desired number of cells).

5. Grow embryoid bodies for 2 days changing growth medium daily or other cells of interest for the appropriate time.

3.2. Immunocytochemistry

All steps are performed at room temperature unless stated otherwise (see Note 2).

1. Wash cells two times for 5 min with PBS.

2. Fix cells for 10 min in 3.7% formaldehyde (Sigma).

3. Wash cells two times for 5 min with PBS.

4. Permeabilize cells in 0.5% Triton-X (Sigma)/PBS for 12 min.

5. Wash cells two times for 5 min with PBS.

6. Block cells for 30–45 min in 1% bovine serum albumin (BSA) in PBS at 37°C.

7. Wash cells two times for 5 min with 1% BSA/PBS.

8. Incubate in primary antibody either overnight at 4° or for 1.5 h at 37° (see Note 3). Dilute antibodies in 1% BSA/PBS.

    (a) Mouse monoclonal anti-GM130 at a 1:50 dilution for Golgi apparatus.

    (b) Mouse monoclonal anti-α-tubulin at a 1:500 dilution for microtubules.

    (c) Mouse monoclonal anti-vinculin at a 1:400 dilution for focal adhesions.

9. Wash cells five times for 5 min in 1% BSA/PBS.

10. Incubate in appropriate Alexa-Fluor (Invitrogen) secondary antibody for 1 h at room temperature covered with foil (see Note 4). Dilute secondary antibodies 1:1,000 in 1% BSA/PBS.

11. Wash cells five times for 5 min in 1% BSA/PBS.

12. Incubate cells in phalloidin for 30 min covered with foil. Dilute phalloidin 1:100 in 1% BSA/PBS. Then proceed with Hoechst if visualizing the nuclei.

13. Wash cells five times for 5 min in 1% BSA/PBS.

14. Incubate cells in Hoechst 33342 diluted at 1:10,000 for 10 min at room temperature covered with foil (see Note 5).

15. Wash cells four times in PBS for 5 min.

16. Mount in gelvatol/NPG by flipping the coverslip onto a superfrost slide with a pea-sized drop of gelvatol/NPG or preferred mounting media.

17. Allow slides to set overnight at RT protected from the light.

3.3. Outgrowth Assay, Extent, and Orientation

3.3.1. Outgrowth Assay

To measure oriented migration, take 4 multichannel fluorescent images of different outgrowths at 20× where the leading edge of the migrating cell sheet is visible. Note the location of the embryoid body in each image. 8.5″ by 11″ images were printed on a color laser-jet printer (Fig. 1).

1. Draw radial lines, approximately 2.5″ apart, from the center of the embryoid body to the leading edge of cells.

2. Divide cells into thirds as shown in Fig. 2 with the front of the "Y" facing the leading edge, with the middle point in the center of the nucleus, and aligning the straight portion of the line with the radial lines from the embryoid body.

3. Score cells as positive for oriented migration when greater than 50% of the Golgi apparatus is localized in the front third of the cell and when greater than 33% of all cells have their Golgi apparatus localized in the front third.

4. Count and score approximately 200 cells per experiment performed in triplicate. Migration is considered to be not oriented when approximately 33% or less of cells have their Golgi apparatus localized in front of the nucleus in the direction of migration.

5. Once percentages have been calculated on triplicates, the averages can be arc-sine transformed by taking the inverse sine of the average for statistics. Standard error of the mean can be done on the arc-sine transformed averages for error bars and a two-tailed student's t-test can be performed on the arc-sine transformed numbers for statistical significance.

Microtubules are observed and cells are considered oriented if they are aligned parallel to the direction of migration.

3.3.2. Migration Extent

Migration extent can be measured two ways: on outgrowths prior to fixing or on Hoechst-stained outgrowths. We have found that it is easier to measure migration on Hoechst-stained outgrowths for transient transfections since the outgrowths need to be grown for 3 days instead of 2 (Subheading 3.5). We have compared measurements using both methods on the same outgrowths and there is no significant difference in migration extent (Fig. 1).

1. Take images of five embryoid bodies using either a 4 or 5× objective to encompass the entire outgrowth in the image.

2. Open images in Adobe Photoshop. Using the measure tool, anchor the tool on the center of the embryoid body and drag towards the top edge of the outgrowth.

3. Record the distance.

4. Keeping the tool anchored on the center, drag the cursor to the edge of the embryoid body.

5. Record the distance.

6. Repeat measurements at 45° intervals around the entire outgrowth for a total of eight measurements.

7. Subtract the center of embryoid body to edge of embryoid body measurement from the center of embryoid body to edge of outgrowth measurements for migration extent for that point.

8. Average all eight migration extent measurements.

9. Average the averages for all five embryoid bodies per condition/ experiment. Once experiments has been performed in triplicate, average the experiments, calculate standard error of the mean and perform a two-tailed student's t-test for statistical significance.

3.4. PCP Inhibitors

3.4.1. sFRP

sFRP1 is added directly to the cell culture medium and binds Wnts, thereby inhibiting them from binding to the Frizzled receptor (Fig. 1).

1. Reconstitute sFRP1 in 1 ml of sterile PBS. The final concentration will be 25 mg/ml and store at −20°C as 50-µl aliquots (see Note 6).

2. Grow embryoid bodies on coverslips for 24 h. After 24 h, change medium, adding only 450 µl of fresh medium instead of 1 ml and add 50 µl sFRP to the medium for a final concentration of 2.5 µg/ml (see Notes).

3. After 24 h, fix and assay cells for oriented migration as described in Subheadings 3.2 and 3.3.

3.4.2. Inhibition at the Level of Rho Kinase

Rho/ROCK signaling can be inhibited at the level of Rho Kinase (ROCK) using the cell-permeable pharmacological inhibitor Y27632, which is highly specific for ROCK.

1. Reconstitute 1 mg/ml Y27632 in 1 ml of PBS and store at −20°C as 50-μl aliquots (see Note 7).

2. Grow embryoid bodies for 24 h on fibronectin-coated coverslips for 24 h. After 24 h, change medium to fresh medium and add 10 μl (10 μM) Y27632 to the medium.

3. After 24 h (48 h after plating), fix and assay cells for oriented migration as described in Subheadings 3.2 and 3.3.

3.4.3. Inhibition Downstream of Rac Signaling

c-jun Kinase (JNK) can modulate signaling downstream of Rac and can be inhibited using the pharmacological inhibitor SP600125 (8).

1. Grow embryoid bodies for 24 h on fibronectin-coated coverslips for 24 h. After 24 h, change medium to fresh medium and add 20 μl (10 μM final concentration) SP600125 to the medium. Bulk up assay for western blotting.

2. After 24 h, fix and assay cells for oriented migration as described in Subheadings 3.2 and 3.3. Confirm effectiveness of SP600125 using western blotting (Subheading 3.6).

3.4.4. Inhibition of GSK-3β Using 6-Bromoindirubin-3′-oxime

To demonstrate that the canonical signaling pathway has no role in PCP, inhibit GSK-3β signaling to increase levels of active β-catenin. Then inhibit PCP with sFRP to determine whether increase in β-catenin can rescue loss of oriented migration.

1. Grow embryoid bodies for 24 h on fibronectin-coated coverslips for 24 h. After 24 h, change medium to 450 μl fresh medium and add 3 μl (3 μM final concentration) BIO inhibitor to the medium. Bulk up assay for western blotting without inhibiting PCP with sFRP (see Note 8).

2. Add 50 μl of 25 mg/ml sFRP.

3. After 24 h, fix and assay cells for oriented migration as described in Subheadings 3.2 and 3.3. Confirm effectiveness of BIO inhibitor using western blotting (Subheading 3.6).

3.5. Transient Transfection

1. Make up 50–100 ml of F9 medium without antibiotics (Subheading 2.1).

2. Plate embryoid bodies in 1 ml of antibiotic-free medium on fibronectin-coated coverslips for immunofluorescence or 10 cm gelatin-coated dishes for western blotting.

3. 35-mm dishes will contain a total of 1 ml of transfection reagent. For the control dish, add 1 ml of antibiotic-free medium.

(a) For each transfection, set up two eppendorf tubes: one for the plasmid and one for the Lipofectamine 2000. Add 100 μl of antibiotic-free medium to each tube.

(b) Add 4 μl of Lipofectamine 2000 to each Lipofectamine-labeled tube and incubate for 5 min at room temperature.

(c) Add 1.6 μg of DNA to plasmid tubes.

(d) After 5 min incubation, add the contents of the DNA tube to the Lipofectamine 2000 tube.

(e) Incubate for 20–30 min at room temperature.

(f) Aspirate old media off dishes.

- Add 1 ml of antibiotic-free medium to control dish.

- For transfection dishes, add 800 μl of antibiotic-free medium to the transfection tube and add the transfection reagent to the appropriate dish.

4. Incubate cells in transfection reagents for 24 h.

5. After 24 h, replace the transfection reagent with fresh F9 medium containing antibiotics and allow outgrowths to recover for 24 h.

6. This protocol can be bulked up for larger experiments. For 10-cm dishes, plate embryoid bodies in 10 ml of antibiotic-free medium and transfect with a total volume of 5 ml.

3.6. Western Blotting Protocol

3.6.1. Collecting Cells

1. For outgrowth cell collection, use transfer pipettes and ice-cold PBS to gently blow the embryoid bodies off the dish. Scrape cells off the dish using a rubber policeman in ice-cold PBS supplemented with protease and phosphatase inhibitors.

2. Centrifuge cells for 15 min at $13,000 \times g$.

3. Aspirate off PBS and snap freeze in dry ice.

3.6.2. Making Cell Lysates

1. Lyse snap frozen cells for 1 h on ice with lysis buffer containing protease and phosphatase inhibitors.

2. Clear lysates by centrifuging for 15 min at $13,000 \times g$.

3. Perform either a Bradford Assay, or use Precision Red to determine Protein concentration. Normalize samples.

4. Remove lysate and put in a new eppendorf tube adding an equal amount of 2× loading buffer.

5. Boil samples for 5 min.

6. Briefly spin samples to remove precipitate.

7. Load up to 30 μl of sample per well in a 10-well gel or 15 μl for a 14-well gel. Load 10 μl of protein standard.

8. Run gel for 1½ h at 125 V in running buffer.

3.6.3. Transfer

1. During last 15 min of gel running, prep Immobilon P membrane.

   (a) Soak membrane for 5 min in methanol.

   (b) Soak membrane for 5 min in distilled water.

2. Equilibrate both membrane and the gel by incubating for 20 min in transfer buffer.

3. Set up transfer apparatus with gel and membrane.

4. Transfer gel in cold room for 1–1½ h at 100 V on a stir plate.

5. When transfer is complete, remove gel from apparatus, rinse briefly in TBS-Tween.

6. Block membrane for 30 min in Blocking Buffer.

   (a) Incubate in secondary antibody for 1 h and detect by ECL for no primary control (Subheading 3.6).

7. Incubate blot overnight rocking in primary antibody diluted in Blocking Buffer at 4°C.

3.6.4. Secondary Antibody and ECL Detection

1. Wash blot five times for 5 min each in TBS-Tween while rocking.

2. Incubate membrane while rocking for 1 h in HRP-conjugated antibody at room temperature.

3. Wash blot five times for 5 min each in TBS-Tween.

4. In a darkroom incubate the membrane in equal volumes of ECL reagents for 1 min. Blot excess with blotting paper.

5. Wrap membrane in plastic wrap or a clear page protector and place in x-ray film cassette. Place a sheet of x-ray film on top of the membrane. Close cassette to expose film. Exposure times:

   (a) Abundant proteins: 1 min or less.

   (b) Moderate protein: 3–5 min.

   (c) Low levels of protein: 10 min is usually the longest that ECL is active for low levels of protein.

6. Develop film in x-ray developing machine.

Stripping Western Blots

1. Incubate blot in stripping buffer with beta-mercaptoethanol (βME) rocking at 50°C for 30 min to 1 h.

2. Wash blot in TBS-Tween while rotating for as many washes as it takes to no longer smell like βME.

3. Block blot in Blotto for 30 min and then reprobe and repeat.

4. Notes

1. We use Tryple Express instead of trypsin as Tryple Express is more stable over time when repeatedly warmed and cooled.

2. We perform immunofluorescence separately for the Golgi apparatus and microtubules using mouse monoclonal antibodies. There is a rabbit polyclonal antibody available for the Golgi apparatus, but there is increased background when performing double immunofluorescence with Golgi and microtubules, making quantification difficult.

3. Incubating overnight at 4°C is preferred to 1.5 h at 37°C because background is increased at 37°C.

4. We prefer Alexa-Fluor 568, but Alexa-Fluor 488 can also be used.

5. Unused Hoechst can be stored at 4°C in a conical tube wrapped in foil for at least 1 month.

6. When using sFRP1, plan out experiments in advance as it has a shelf life of approximately 4–6 months when stored at –20°C after it has been reconstituted. We tested both 2.5 and 5 µg/ml and did not find any significant difference between the higher and lower dose, therefore we continued experiments with the lower dose.

7. We have found that we can store reconstituted aliquots of Y27632 at –20°C for approximately 1 year without loss of effectiveness.

8. We tested 1, 2, and 3 µM and did not see any effect on orientation. Western blotting was performed on controls and BIO-treated dishes only to show that levels of active β-catenin were increased in response to BIO treatment.

References

1. Klein, T. J., and Mlodzik, M. (2005) Planar Cell Polarization: An Emerging Model Points in the Right Direction. *Annu Rev Cell Dev Biol 21*, 155–76.

2. Kiefer, J. C. (2005) Planar cell polarity: heading in the right direction. *Dev Dyn 233*, 695–700.

3. Keller, R. (2002) Shaping the vertebrate body plan by polarized embryonic cell movements. *Science 298*, 1950–4.

4. Montcouquiol, M., Crenshaw, E. B., 3rd, and Kelley, M. W. (2006) Noncanonical Wnt signaling and neural polarity. *Annu Rev Neurosci 29*, 363–86.

5. De Calisto, J., Araya, C., Marchant, L., Riaz, C. F., and Mayor, R. (2005) Essential role of non-canonical Wnt signalling in neural crest migration. *Development 132*, 2587–97.

6. Yamanaka, Y., Ralston, A., Stephenson, R. O., and Rossant, J. (2006) Cell and molecular regulation of the mouse blastocyst. *Dev Dyn 235*, 2301–14.

7. Grabel, L. B., and Watts, T. D. (1987) The role of extracellular matrix in the migration and differentiation of parietal endoderm from teratocarcinoma embryoid bodies. *J Cell Biol 105*, 441–8.

8. Rosso, S. B., Sussman, D., Wynshaw-Boris, A., and Salinas, P. C. (2005) Wnt signaling through

Dishevelled, Rac and JNK regulates dendritic development. *Nat Neurosci 8*, 34–42.

9. Kishida, S., Yamamoto, H., and Kikuchi, A. (2004) Wnt-3a and Dvl induce neurite retraction by activating Rho-associated kinase. *Mol Cell Biol 24*, 4487–501.

10. Coso, O. A., Chiariello, M., Yu, J. C., Teramoto, H., Crespo, P., Xu, N., Miki, T., and Gutkind, J. S. (1995) The small GTP-binding proteins Rac1 and Cdc42 regulate the activity of the JNK/SAPK signaling pathway. *Cell 81*, 1137–46.

11. Yamanaka, H., Moriguchi, T., Masuyama, N., Kusakabe, M., Hanafusa, H., Takada, R., Takada, S., and Nishida, E. (2002) JNK functions in the non-canonical Wnt pathway to regulate convergent extension movements in vertebrates. *EMBO Rep 3*, 69–75.

12. Boutros, M., Paricio, N., Strutt, D. I., and Mlodzik, M. (1998) Dishevelled activates JNK and discriminates between JNK pathways in planar polarity and wingless signaling. *Cell 94*, 109–18.

13. Mills, E., LaMonica, K., Hong, T., Pagliaruli, T., Mulrooney, J., and Grabel, L. (2005) Roles for Rho/ROCK and vinculin in parietal endoderm migration. *Cell Commun Adhes 12*, 9–22.

14. LaMonica, K., Bass, M., and Grabel, L. (2009) The planar cell polarity pathway directs parietal endoderm migration. *Dev Biol 330*, 44–53.

15. Chen, B. H., Tzen, J. T., Bresnick, A. R., and Chen, H. C. (2002) Roles of Rho-associated kinase and myosin light chain kinase in morphological and migratory defects of focal adhesion kinase-null cells. *J Biol Chem 277*, 33857–63.

16. Habas, R., Kato, Y., and He, X. (2001) Wnt/Frizzled activation of Rho regulates vertebrate gastrulation and requires a novel Formin homology protein Daam1. *Cell 107*, 843–54.

17. Sato, A., Khadka, D. K., Liu, W., Bharti, R., Runnels, L. W., Dawid, I. B., and Habas, R. (2006) Profilin is an effector for Daam1 in non-canonical Wnt signaling and is required for vertebrate gastrulation. *Development 133*, 4219–31.

18. Sato, N., Meijer, L., Skaltsounis, L., Greengard, P., and Brivanlou, A. H. (2004) Maintenance of pluripotency in human and mouse embryonic stem cells through activation of Wnt signaling by a pharmacological GSK-3-specific inhibitor. *Nat Med 10*, 55–63.

19. Worthylake, R. A., Lemoine, S., Watson, J. M., and Burridge, K. (2001) RhoA is required for monocyte tail retraction during transendothelial migration. *J Cell Biol 154*, 147–60.

20. Worthylake, R. A., and Burridge, K. (2003) RhoA and ROCK promote migration by limiting membrane protrusions. *J Biol Chem 278*, 13578–84.

Chapter 16

Analysis of Wnt/Planar Cell Polarity Pathway in Cultured Cells

Mitsuharu Endo, Michiru Nishita, and Yasuhiro Minami

Abstract

Planar cell polarity (PCP) pathway of Wnt signaling plays a crucial role to establish the polarization of cells during tissue development. Our recent findings using in vitro analyses have revealed that Ror2, a member of the Ror-family receptor tyrosine kinases, acts as a receptor or co-receptor for Wnt5a and plays a crucial role for Wnt5a-induced polarized cell migration through activating PCP pathway. Indeed, analyses of both *Wnt5a* and *Ror2* mutant mice have shown that Wnt5a-Ror2 signaling is involved in establishing the PCP in epithelial tissues in vivo, indicating that in vitro analyses of polarized cell migration and PCP signaling induced by Wnt5a can be useful tools to explore putative regulators involved in Wnt/PCP pathway. Here, we introduce in vitro methods using cultured cells to monitor polarized cell migration and PCP signaling induced by Wnt5a.

Key words: PCP, Wnt5a, Ror2, JNK, Non-canonical Wnt signaling, Cell polarity, Wound-healing

1. Introduction

Planar cell polarity (PCP) is characterized as the coordinated orientation of cells within the plane of epithelial tissues in vertebrates and invertebrates. A set of genes, involved in regulating the PCP, have been identified from the genetic studies in *Drosophila* (1–3). In vertebrates, a similar group of genes, including *frizzled* and *dishevelled*, involved in the regulation of Wnt signaling, have been shown to regulate the establishment of the PCP (4, 5). Vertebrate Wnt ligands can be classified into at least two subfamilies; one is known as the Wnt1 class and signals primarily through the canonical Wnt/β-catenin pathway (6, 7). The other is known as the Wnt5a class and signals mainly through the β-catenin independent non-canonical Wnt pathways, including the PCP pathway (8–12).

Kursad Turksen (ed.), *Planar Cell Polarity: Methods and Protocols*, Methods in Molecular Biology, vol. 839,
DOI 10.1007/978-1-61779-510-7_16, © Springer Science+Business Media, LLC 2012

It has recently been shown that Wnt5a functions to coordinate the uniform orientation of stereociliary bundles of sensory hair cells, which is a representative epithelial PCP observed in the inner-ear sensory organ of mice (13). However, the molecular mechanism underlying Wnt/PCP pathway is not fully understood yet.

We examined by using in vitro wound-healing assay how polarization signals are transmitted to cell interior following Wnt5a stimulation. We found that the receptor tyrosine kinase Ror2 plays critical roles in Wnt5a-induced polarized cell migration (14). Furthermore, we found that Wnt5a stimulation induces activation of the c-Jun N-terminal kinase (JNK), a critical regulator of the PCP pathway in both invertebrates and vertebrates (15, 16), in cells at the wound-edge in a Ror2-dependent manner (14). It has been well established that activation of JNK occurs through phosphorylation at both threonine 183 and tyrosine 185 in it (17). Thus, the phosphorylation sites-specific antibody that recognizes this phosphorylated form of JNK can be used to monitor the activity of JNK by Western blotting. As an alternative approach to monitor activity of JNK, we employed an AP-1-driven luciferase reporter analysis. This analysis can be used as a powerful tool to monitor the activation of JNK pathway with high sensitivity and reliability (18, 19). In fact, we have shown that Wnt5a stimulation induces AP-1 activation in a manner dependent on the expression of Ror2 (20).

Consistent with the findings that Ror2 mediates Wnt5a-induced polarization signals in cultured cells, Ror2 has been implicated in the PCP establishment of sensory hair cells in the inner-ear sensory organ of mice (21). PCP is a higher-order complex phenomenon seen in epithelial tissues in vivo. Therefore, in vitro analyses using cultured cells will provide powerful tools to study Wnt/PCP pathway at cellular levels. The methods described here are also of use to explore candidate molecules involved in Wnt/PCP pathway, although their roles in the PCP establishment have to be assessed in vivo.

2. Materials

2.1. Cell Culture

1. Dulbecco's Modified Eagle's Medium (DMEM, Nissui) supplemented with 5–10% fetal bovine serum (FBS, Biowest).

2. Phosphate buffered saline (PBS): 10× stock: 1.37 M NaCl, 27 mM KCl, 100 mM Na_2HPO_4, 18 mM KH_2PO_4 (adjust to pH 7.4 with HCl), autoclaved prior to its storage at room temperature. Prepare working solution by diluting 10× stock with water.

3. Trypsin (2.5%, GIBCO): Working solution (0.25%) is prepared by diluting with PBS and stored at 4°C.

4. Fibronectin (Sigma): Working solution (10 µg/ml) is prepared by diluting with PBS just prior to use.

5. Purified mouse Wnt5a (R&D Systems) is dissolved at 20 µg/ml in PBS containing 0.1% (w/v) fraction V bovine serum albumin (BSA, Wako) and stored in single use aliquots at −80°C. Working solutions are prepared by dilution with DMEM containing 0.1% (w/v) BSA.

6. GeneSilencer siRNA Transfection Reagent (Gene Therapy Systems).

7. Lipofectamine 2000 Reagent (Invitrogen).

8. siRNA: The mouse *Ror2* siRNA #1 (AAGAUUCGGAG-GCAAUCGACA) and control siRNA (GUACCGCAC-GUCAUUCGUAUC) from RNAi Co., Ltd. The mouse *Ror2* siRNA #2 (UCCCAUCCUUCUGCCACUUCGUCUU) from Invitrogen.

2.2. Cell Lysis and SDS-Polyacrylamide Gel Electrophoresis

1. Cell lysis buffer: 50 mM Tris–HCl (pH 7.4), 150 mM NaCl, 0.5% (v/v) Nonidet P-40, 5 mM EDTA, 50 mM NaF, 1 mM Na_3VO_4, 1 mM phenylmethylsulfonyl fluoride (PMSF), 10 µg/ml leupeptin, 10 µg/ml aprotinin.

2. Laemmli's SDS sample buffer (5×): 125 mM Tris–HCl (pH 6.8), 5% SDS, 50% (v/v) glycerol, 5% β-mercaptoethanol, 0.1% bromophenol blue. Store in aliquots at −20°C.

3. Running buffer (5×): 125 mM Tris, 960 mM glycine, 0.5% SDS. Store at room temperature.

4. Prestained molecular weight markers: Precision plus protein standards (Bio-Rad).

2.3. Western Blotting

1. Transfer buffer: 25 mM Tris, 190 mM glycine, 15% (v/v) methanol. Store at room temperature.

2. PVDF membrane: Immobilon-P transfer membrane (Millipore).

3. PBS-T: PBS containing 0.05% (v/v) Tween-20.

4. Blocking buffer for detection of non-phosphorylated proteins: 5% (w/v) non-fat dry milk in PBS-T.

5. Blocking buffer for detection of phosphorylated proteins: 5% (w/v) BSA in PBS-T.

6. Primary antibodies: Anti-Ror2 (22), anti-β-actin (AC-15, Sigma), anti-phospho-JNK (Thr-183/Thy-185, Cell Signaling), and anti-JNK (FL, Santa Cruz Biotechnology).

7. Secondary antibodies: Anti-rabbit IgG antibody conjugated with horseradish peroxidase (HRP) (Bio-Rad), and anti-rabbit IgG antibody conjugated with HRP (Bio-Rad).

8. Enhanced chemiluminescence (ECL) reagents: Western lightning plus-ECL (PerkinElmer).

9. Stripping buffer: 62.5 mM Tris–HCl (pH 6.8), 2% SDS. Store at room temperature. Warm up to 60°C and add 50 mM β-mercaptoethanol prior to use.

**2.4. Immunofluo-
rescence**

1. Circular microscope coverslips (18 mm diameter) from Fisher.
2. 4% Paraformaldehyde phosphate buffer solution from Wako. Store at 4°C.
3. Permeabilization solution: 0.2% (v/v) Triton X-100 in PBS.
4. Blocking and antibody dilution buffer: 5% (w/v) BSA in PBS.
5. Rhodamine–phalloidin (Invitrogen).
6. Primary antibodies: Anti-γ-tubulin (GTU-88, Sigma) and anti-phospho-c-Jun (KM-1, Santa Cruz Biotechnology).
7. Secondary antibody: Anti-mouse IgG antibody conjugated with Alexa Fluor 488 (Invitrogen).
8. Nuclear stain: 4′,6-diamidino-2-phenylindole (DAPI) is dissolved at 1 mg/ml in water.
9. Mounting medium: Prolong antifade (Invitrogen).
10. Microscope glass slide (Matsunami).

**2.5. Luciferase
Reporter Analysis**

1. Renilla luciferase control vector (pGL4.74, Promega).
2. Dual-Luciferase Reporter Assay System (Promega).

3. Methods

Polarized migrating cells possess a leading edge that is a highly motile structure in the direction of forward movement (see Fig. 1a). At the leading edge, cells protrude lamellipodia and filopodia that are driven by polymerization of actin. An in vitro wound-healing analysis is a useful tool to examine the mechanisms of cell polarization. Lamellipodia formation at the wound-edge and reorientation of microtubule organizing center (MTOC) toward the direction of cell migration are observed during the initial step of cell polarization after wounding (23, 24). Because these polarized cells at the wound-edge continue to move toward wounded space, the rate of wound closure after wounding, which is calculated by relative migrating distance between the wound-edges, reflects the extent of cells to acquire the polarity (see Fig. 1b).

The activities of JNK, a crucial component of Wnt/PCP pathway, can be determined by Western blotting using an anti-phospho-JNK antibody that recognizes an active form of JNK. JNK activation can also be assessed by measuring the transcriptional activities of AP-1-family, containing c-Jun, a downstream effector of JNK. The activities of AP-1 can be readily measured by the luciferase reporter analysis where a reporter gene containing several tandem repeats of an AP-1-binding consensus sequence (TGACTAA) at the upstream of a luciferase gene was transfected to cells of interest. It is also important to confirm whether the activation of JNK is

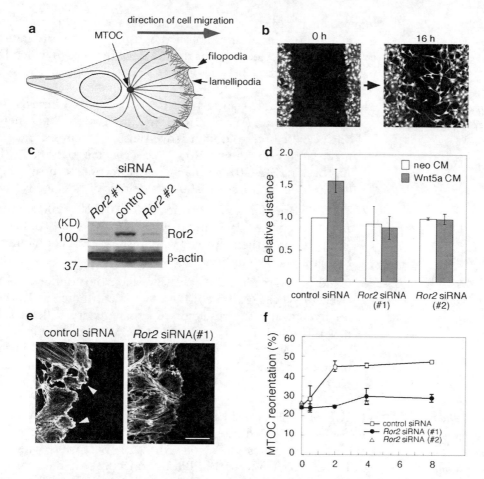

Fig. 1. Ror2 is required for Wnt5a-induced polarized cell migration. (**a**) A diagram of polarized migrating cell. Polarized migrating cells exhibit actin-rich protrusions, such as lamellipodia and filopodia, and reorientation of the microtubule organizing center (MTOC) to a location in front of the nucleus, toward the direction of cell migration. (**b**) Representative images of wound-healing analysis. Confluent monolayers of NIH3T3 cells were wounded with a pipette tip and treated with Wnt5a conditioned medium (CM) for 0 h (*left*) or 16 h (*right*). (**c**) siRNA-mediated suppression of Ror2 expression. NIH3T3 cells were transfected with either control siRNA or *Ror2* siRNA (*Ror2* #1 or #2). After 72 h in culture, whole cell lysates were analyzed by immunoblotting with antibodies against Ror2 and β-actin, respectively. (**d**) Confluent monolayers of cells transfected with either control siRNA or *Ror2* siRNA (*Ror2* #1 or #2) were wounded and treated with control (neo) CM or Wnt5a CM for 16 h. The relative migrating distance of the wound-edge was shown as mean ± SD of three independent experiments, with the migrating distance of control siRNA- and neo CM-treated cells set as 1.0. Suppression of Ror2 expression results in a drastic inhibition of Wnt5a-induced wound closure. (**e**) Cells transfected with either control siRNA or *Ror2* siRNA (*Ror2* #1) were wounded and incubated with Wnt5a CM for 30 min. Cells were fixed and stained with rhodamine–phalloidin to visualize F-actin. *Arrowheads* indicate lamellipodial protrusions at the wound-edge of control siRNA-treated cells. Ror2-depleted cells fail to exhibit lamellipodia formation following Wnt5a stimulation. Bar = 50 μm. (**f**) Cells transfected with either control siRNA or *Ror2* siRNA (*Ror2* #1 or #2) were wounded and treated with Wnt5a CM for the indicated times. The percentage of wound-edge cells with reoriented MTOC was determined. The data are expressed as mean ± SD of three independent experiments. *n* > 100 for each time point in each experiment. Wnt5a-induced MTOC reorientation is impaired in Ror2-depleted cells. (This research (**c–f**) was originally published in the Journal of Biological Chemistry. Nomachi, A., Nishita, M., Inaba, D., Enomoto, M., Hamasaki, M., and Minami, Y. Receptor tyrosine kinase Ror2 mediates Wnt5a-induced polarized cell migration by activating c-Jun N-terminal kinase via actin-binding protein filamin A. *J Biol Chem* **283**, 27973–27981 (2008). © the American Society for Biochemistry and Molecular Biology).

induced in polarized cells. An antibody against phosphorylated c-Jun at Ser-63, the well-established phosphorylation site in c-Jun by activated JNK, clearly stains the nuclei of the cells in which JNK pathway is activated (25).

3.1. Preparation of Active Wnt5a Protein

1. L cell lines stably transfected with *Wnt5a* cDNA inserted into a pPKGneo vector (Wnt5a/L cells) or with an empty vector (neo/L cells) can be used to obtain active Wnt5a protein secreted into their culture medium (26). The cells are passaged when approaching 30–50% confluence with trypsin to provide new maintenance cultures on 100-mm culture dishes (see Note 1).

2. The cells are plated at a density of 5×10^5 cells/100-mm dish and cultured for 2 days until they reach to 80–100% confluence. The medium is then replaced with fresh DMEM containing 5% FBS and further cultured for 24 h.

3. Conditioned media (CM) are harvested from confluent monolayers of Wnt5a/L or neo/L cells and centrifuged at $1,500 \times g$ for 5 min, and the resultant supernatants are collected into new tubes and stored at –80°C. The CM are diluted in serum-free DMEM (1:2.5–1:10) prior to use (see Note 2). Purified mouse (or human) Wnt5a is also available from commercial sources.

3.2. Preparation of Monolayered Cells for In Vitro Wound-Healing Analysis

1. NIH3T3 cells are maintained on 100-mm culture dishes in DMEM containing 10% FBS. The cells are plated onto fibronectin-coated coverslips (see Note 3) in 12-well plates with 1 ml DMEM containing 10% FBS. Fibronectin-coated coverslips in 12-well plates should be prepared prior to use by adding fibronectin solution (10 μg/ml in PBS), incubated for 2 h at room temperature, and then washed twice with PBS. The plated cells must be 50–70% confluent when they are transfected.

2. The cells are transfected with siRNAs using GeneSilencer siRNA Transfection Reagent according to the manufacturer's protocol (Gene Therapy Systems) with a final (siRNA) concentration of 80 nM. Four hours after transfection, medium is replaced with fresh DMEM containing 10% FBS and the cells are grown to become confluent for 24 h.

3.3. Western Blotting for Ror2

1. Control or *Ror2* siRNA (*Ror2* #1 and #2) are transfected into NIH3T3 cells on fibronectin-coated 12-well plates as described above (see Subheading 3.2). Then the cells are cultured for 72 h until the expression of Ror2 are suppressed by *Ror2* siRNA.

2. The cells are washed once with ice-cold PBS, and lysed with 100 μl ice-cold cell lysis buffer on ice for 20 min.

3. The cell lysates are centrifuged at $12,000 \times g$ for 15 min at 4°C to remove insoluble materials. The supernatants are collected

into new microtubes, mixed with Laemmli's sample buffer (5×), and then boiled for 10 min at 95°C. After cooling down to room temperature, 20 µl from the respective samples are applied onto and separated by SDS-polyacrylamide gel electrophoresis (SDS-PAGE) using a 10% polyacrylamide gel that can be purchased as ready-made or prepared in the laboratory (see other general laboratory handbooks to prepare the gels).

4. The separated proteins are transferred onto a PVDF membrane electrophoretically.

5. The PVDF membrane is then incubated in 50 ml blocking buffer [PBS-T containing 5% (w/v) non-fat dry milk] for 1 h at room temperature on a rocking platform.

6. The blocking buffer is removed (see Note 4), and the membrane is incubated with a 1:1,000 dilution of an anti-Ror2 antibody (see Note 5) in PBS-T/1% non-fat dry milk for 2 h at room temperature on a rocking platform.

7. The membrane is then washed three times with PBS-T (10 min for each wash).

8. The secondary antibody (anti-rabbit IgG antibody conjugated with HRP) is freshly prepared for each experiment as a 1:10,000 dilution in PBS-T/1% non-fat dry milk. The membrane is incubated in the above solution containing secondary antibody for 1 h at room temperature on a rocking platform.

9. The membrane is then washed three times with PBS-T (10 min for each wash).

10. Equal amounts of Enhanced Luminol Reagent and Oxidizing Reagent of ECL reagents are mixed just prior to their use. After the final wash, the membrane is incubated with the mixture of ECL reagents for 1 min on plastic wrap.

11. The mixture of ECL reagents is removed from the membrane, and then placed between the leaves of a plastic wrap. The wrap containing the membrane is then placed in an X-ray film cassette with a film for a suitable exposure time. An example of the results is shown in Fig. 1c.

12. The membrane is then incubated with 0.02% sodium azide in PBS-T for 30 min at room temperature on a rocking platform to inactivate the HRP activity.

13. After the inactivation of HRP, the membrane is washed three times with PBS-T (5 min for each wash), and then incubated again in blocking buffer (see Note 4).

14. The membrane is then reprobed with an anti-β-actin antibody (1:10,000 in PBS-T/1% non-fat dry milk) as a loading control, followed with secondary antibody (anti-mouse IgG antibody conjugated with HRP), and detected by ECL as described above. An example of the results is shown in Fig. 1c.

**3.4. In Vitro
Wound-Healing
Analysis**

1. Confluent monolayers of cells transfected with control siRNA or *Ror2* siRNA (*Ror2* #1 and #2) are prepared as described above (see Subheading 3.2).

2. The monolayers are wounded with a pipette tip once across the coverslip and washed twice with PBS to remove cell debris (see Note 6). The cells were cultured with control (neo) CM or Wnt5a CM for 16 h.

3. The cells are then fixed with 4% paraformaldehyde in PBS for 15 min at room temperature and washed three times with PBS.

4. The coverslips are carefully inverted onto a drop (10–20 μl) of PBS on a microscope slide and sealed with nail varnish.

5. The slides are viewed under phase contrast microscopy and the widths of wounds are measured to calculate migrating distance of each wound-edge. The relative migrating distance of the wound-edge is determined by subtracting the width of wound at 16 h from that at 0 h (see Fig. 1b). The migrating distance of control cells (control siRNA- and neo CM-treated cells in the experiment) is set as 1.0. An example of the results is shown in Fig. 1d.

**3.5. Lamellipodia
Staining at
Wound-Edge**

1. The wounded cells transfected with control siRNA or *Ror2* siRNA (*Ror2* #1) are prepared as described above (see Subheading 3.4). The cells are treated with neo CM or Wnt5a CM for 30 min.

2. The cells are then fixed with 4% paraformaldehyde in PBS for 15 min at room temperature and washed three times with PBS.

3. The cells are permeabilized by incubation with 0.2% Triton X-100 in PBS for 5 min at room temperature and then washed twice with PBS.

4. For blocking, the samples are incubated with 5% BSA in PBS for 1 h at room temperature.

5. After blocking, the coverslips are incubated with a 1:1,000 dilution of rhodamine–phalloidin in PBS/5% BSA (see Note 7) for 1 h at room temperature to stain the filamentous actin (F-actin).

6. The samples are then washed three times with PBS (10 min for each wash).

7. The coverslips are carefully inverted onto a drop (10–20 μl) of mounting medium on a microscope slide and sealed with nail varnish. The sample can be viewed immediately when varnish is dry, or stored in dark at 4°C for up to a month.

8. The slides are viewed under confocal laser microscopy. Excitation at 543 nm induces the rhodamine fluorescence. An example of the images for F-actin staining is shown in Fig. 1e.

**3.6. MTOC
Reorientation Analysis**

1. The wounded cells transfected with control siRNA or *Ror2* siRNA (*Ror2* #1) are prepared as described above (see Subheading 3.4). The wounded cells are immediately fixed (0 min) or treated with neo CM or Wnt5a CM for 30 min, 2, 4, or 8 h, then fixed with 100% methanol for 5 min at −20°C (see Note 8), and washed three times with PBS.

2. For blocking, the samples are incubated with 5% BSA in PBS for 1 h at room temperature.

3. After blocking, the coverslips are incubated with a 1:1,000 dilution of an anti-γ-tubulin antibody in PBS/5% BSA (see Note 7) for 2 h at room temperature or overnight at 4°C to stain the MTOC.

4. The samples are then washed three times with PBS (10 min for each wash), and then incubated with a 1:1,000 dilution of the secondary antibody (anti-mouse IgG antibody conjugated with Alexa Fluor 488) and a 1:1,000 dilution of DAPI in PBS/5% BSA for 1 h at room temperature.

5. After washing, the coverslips are mounted on microscope slides and viewed under confocal laser microscopy as described above (see Subheading 3.5). Excitation at 488 nm induces the Alexa Fluor 488 fluorescence for the γ-tubulin (MTOC), while excitation at 360 nm induces DAPI fluorescence for the nuclei. The rate of wound-edge cells in which the MTOC is within the 90° sector from the center of nuclei to the wound-edge is determined as the rate of MTOC reorientation (see Note 9). An example of the results analyzing the rate of MTOC reorientation is shown in Fig. 1f.

**3.7. Detection
of Active JNK
by Western Blotting**

1. Confluent monolayers of NIH3T3 cells transfected with control siRNA or *Ror2* siRNA (*Ror2* #1) are prepared on fibronectin-coated dishes (35 mm in diameter) as described above (see Subheading 3.2). The media are replaced with DMEM containing 1% FBS, and 24 h after culturing, the monolayers are scratched eight times across the dish.

2. Immediately after wounding, the cells are stimulated with Wnt5a by adding purified Wnt5a (final conc. of 200 ng/ml) or its vehicle (PBS containing 0.1% BSA) for 20 min (see Note 10).

3. The cell lysates from the respective samples are prepared and separated by SDS-PAGE as described above (see Subheading 3.3). The separated proteins are transferred onto a PVDF membrane electrophoretically, and the membrane is incubated with blocking buffer (see Note 11) for 1 h at room temperature on a rocking platform.

4. After blocking, the membrane is probed with an anti-phospho-JNK antibody (1:1,000 in PBS-T/1% BSA—see Note 12) followed with the secondary antibody (anti-rabbit IgG anti-

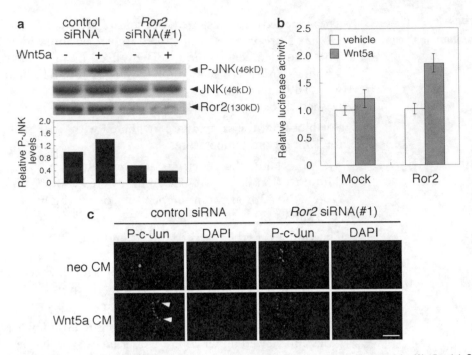

Fig. 2. Ror2 is required for JNK activation at the wound-edge and AP-1 activation induced by Wnt5a. (**a**) Confluent monolayers of NIH3T3 cells transfected with control siRNA or *Ror2* siRNA (*Ror2* #1) were wounded and treated with purified Wnt5a (200 ng/ml) or its vehicle for 20 min. The levels of Ror2, P-JNK, and total JNK were analyzed by immunoblotting, respectively. The histograms indicate the relative P-JNK levels. JNK phosphorylation induced by Wnt5a was remarkably inhibited in Ror2-depleated cells. (**b**) L cells were transfected with mock or *Ror2* expression plasmids along with *AP-1* reporter plasmids. After serum starvation for 12 h, the cells were treated with either Wnt5a (400 ng/ml) or its vehicle for 24 h and subjected to luciferase assays. Wnt5a stimulation dominantly induces transcriptional activation of an AP-1-Luc in Ror2-expressing cells. (**c**) Confluent monolayers of NIH3T3 cells transfected with control siRNA or *Ror2* siRNA (*Ror2* #1) were wounded and treated with neo CM or Wnt5a CM for 30 min. Cells were costained with anti-phospho-c-Jun antibody (P-c-Jun) and DAPI. *Arrowheads* indicate the wound-edge cells with enhanced P-c-Jun staining after Wnt5a treatment. Wnt5a-induced c-Jun phosphorylation in wound-edge cells is impaired in Ror2-depleted cells. Bar = 100 μm. (This research (**a, c**) was originally published in the Journal of Biological Chemistry. Nomachi, A., Nishita, M., Inaba, D., Enomoto, M., Hamasaki, M., and Minami, Y. Receptor tyrosine kinase Ror2 mediates Wnt5a-induced polarized cell migration by activating c-Jun N-terminal kinase via actin-binding protein filamin A. *J Biol Chem* **283**, 27973–27981. (2008) © the American Society for Biochemistry and Molecular Biology (**b**, unpublished data)).

body conjugated with HRP) and detected by ECL as described above (see Subheading 3.3). An example of the results is shown in Fig. 2a.

5. The membrane is then incubated with the stripping buffer for 30 min at 60°C with occasional agitation.

6. After stripping off previous signals, the membrane is washed three times with PBS-T (5 min for each wash) and then incubated in blocking buffer.

7. The membrane is then reprobed with an anti-(total) JNK antibody (1:1,000 in PBS-T/1% BSA) that recognizes JNK irrespective of its phosphorylation status, followed with the secondary

antibody (anti-rabbit IgG antibody conjugated with HRP), and detected by ECL as described above. This provides a loading control, confirming comparable recovery from the different samples throughout the procedure. An example of the result is shown in Fig. 2a.

3.8. Luciferase Reporter Analysis for AP-1 Activity

1. L cells are maintained on 100-mm culture dishes in DMEM containing 5% FBS. The cells are plated onto fibronectin-coated 48-well plates with 200 μl DMEM containing 5% FBS. The plated cells must be 50–70% confluent when they are transfected.

2. The cells are transfected with mock or *Ror2* expression plasmids (20) along with *AP-1* reporter plasmids (20) using Lipofectamine 2000 according to the manufacturer's protocol (Invitrogen). Amounts of DNA transfected are 61.25 ng pAP-1-Luc, 1.25 ng phRluc/TK (to normalize for transfection efficiency), and 62.5 ng pEGFP or pEGFP-Ror2. Three hours after transfection, the media are replaced with fresh DMEM containing 5% FBS and the cells are grown for 6 h.

3. Transfected cells are then serum-starved with DMEM containing 0.1% BSA for 12 h and then treated with purified Wnt5a (final conc. of 200 ng/ml) or its vehicle (PBS containing 0.1% BSA) for 24 h.

4. Luciferase activities (Firefly luciferase expressed from pAP-1-Luc and Renilla luciferase expressed from phRluc/TK) are determined by using the Dual-Luciferase Reporter Assay System (Promega) and Microplate Luminometer. The relative luciferase activities are determined as the ratio of Firefly to Renilla, and the activity of control cells (mock-transfected and vehicle-treated cells in the experiment) is set as 1.0. An example of the results is shown in Fig. 2b.

3.9. Detection of Phospho-c-Jun at Wound-Edge by Immunofluorescence

1. The wounded cells transfected with control siRNA or *Ror2* siRNA (*Ror2* #1) are prepared as described above (see Subheading 3.4). The cells are treated with neo CM or Wnt5a CM for 30 min.

2. Cell fixation and blocking procedures are performed as described above (see Subheading 3.6).

3. After blocking, the coverslips are incubated with a 1:1,000 dilution of an anti-phospho-c-Jun (P-c-Jun) antibody in PBS/5% BSA (see Note 7) for 2 h at room temperature or overnight at 4°C.

4. The samples are then washed three times with PBS (10 min for each wash), and then incubated with a 1:1,000 dilution of the secondary antibody (anti-mouse IgG antibody conjugated with Alexa Fluor 488) and a 1:1,000 dilution of DAPI in PBS/5% BSA for 1 h at room temperature.

5. After washing, the coverslips are mounted on microscope slides and viewed under confocal laser microscopy as described above (see Subheading 3.5). Excitation at 488 nm induces the Alexa Fluor 488 fluorescence for P-c-Jun, while excitation at 360 nm induces DAPI fluorescence for the nuclei. An example of the results is shown in Fig. 2c.

4. Notes

1. The passage of cells should be done during the log-phase growth. When the cells reach to 80–100% confluence, the morphologically large cells are appeared and increased, which leads to inefficient production of Wnt5a.

2. Because lipids, such as lysophosphatidic acid, and growth factors, contained in the serum, might affect the basal cellular responses (e.g. migration, MTOC reorientation, or JNK activation) following scratching the monolayers (irrespective of the presence of Wnt5a), CM with low serum concentration (0.5–2%) should be used.

3. Coating of coverslips with fibronectin seems to be necessary to measure the activities of JNK following Wnt5a stimulation of NIH3T3 cells, because Wnt5a-induced JNK activation is hardly detectable in cells cultured on non-coated dishes.

4. The blocking buffer can be saved for subsequent uses, including the dilution of primary and secondary antibodies and the blocking of membrane after stripping, by storing at 4°C, although prolonged storage reduces its blocking capacity.

5. We use a rabbit polyclonal anti-mouse Ror2 antibody that was previously prepared in our laboratory (22). Anti-Ror2 antibodies are also available from several commercial sources, although we have not checked whether the commercial antibodies can be used for Western blotting or other methods.

6. It is important to make a constant-width wound for each sample. To calculate the migrating distance of wound-edge after culturing, it is necessary to measure the width of the wound in one sample immediately after scratching (it is considered at 0 h).

7. Only 50–75 μl (or 100–150 μl if you use 22 mm x 22 mm coverslips in 6-well plates) of diluted rhodamine–phalloidin (or antibodies) solution per sample can be used at this step. To this end, coverslips are carefully inverted onto a drop of diluted solution on a sheet of Parafilm to save the amount of the solution and to prevent drying up.

8. Methanol fixation seems to be preferable to preserve the structures of the microtubules and MTOC. An anti-γ-tubulin antibody (and an anti-phospho-c-Jun antibody) utilized in our experiments can recognize γ-tubulin proteins (and phospho-c-Jun proteins) even in cells fixed with methanol, although methanol fixation sometimes destroys antigenic epitopes by denaturing proteins. Because methanol also permeabilizes cell membranes, a step of permeabilization with detergents such as Triton X-100 can be omitted.

9. Because non-polarized cells show random orientation of the MTOC around their nuclei, the rate of MTOC reorientation in the cells is around 25%.

10. Essentially identical results are obtained using Wnt5a CM under optimized conditions (14).

11. Blocking agents, which may contain phosphoproteins and/or protein phosphatases such as PBS-T/5% non-fat dry milk, are not well suited for immunoblotting with anti-phospho-specific antibodies. The signals of phosphorylated-JNK are clearly detected with low backgrounds by using PBS-T/5% BSA as a blocking solution.

12. The sequences surrounding activating phosphorylation sites are somewhat conserved between JNK and Erk. Although the anti-phospho-JNK antibody is highly specific to phosphorylated-JNK, this antibody seems to recognize phosphorylated-Erk as well. The bands of phospho-Erk (42 and 44 kDa) exhibit slightly higher mobilities than those of phospho-JNK (46 and 54 kDa). To confirm the identity of the detected bands, it is important that the same membrane is reprobed with an anti-(total) JNK antibody and the molecular sizes are compared between the bands detected by anti-phospho-JNK and anti-(total) JNK antibodies.

References

1. Gubb, D., and Garcia-Bellido, A. (1982) A genetic analysis of the determination of cuticular polarity during development in Drosophila melanogaster. *J Embryol Exp Morphol* **68**, 37–57.

2. Adler, P. N. (2002) Planar signaling and morphogenesis in Drosophila. *Dev Cell* **2**, 525–535.

3. Klein, T. J., and Mlodzik, M. (2005) Planar cell polarization: an emerging model points in the right direction. *Annu Rev Cell Dev Biol* **21**, 155–176.

4. Mlodzik, M. (2002) Planar cell polarization: do the same mechanisms regulate Drosophila tissue polarity and vertebrate gastrulation? *Trends Genet* **18**, 564–571.

5. Fanto, M., and McNeill, H. (2004) Planar polarity from flies to vertebrates. *J Cell Sci* **117**, 527–533.

6. Cadigan, K. M., and Nusse, R. (1997) Wnt signaling: a common theme in animal development. *Genes Dev* **11**, 3286–3305.

7. Sokol, S. Y. (1999) Wnt signaling and dorsoventral axis specification in vertebrates. *Curr Opin Genet Dev* **9**, 405–410.

8. Moon, R. T., Campbell, R. M., Christian, J. L., McGrew, L. L., Shih, J., and Fraser, S. (1993) Xwnt-5A: a maternal Wnt that affects morphogenetic movements after overexpression in embryos of Xenopus laevis. *Development* **119**, 97–111.

9. Heisenberg, C. P., Tada, M., Rauch, G. J., Saude, L., Concha, M. L., Geisler, R., Stemple, D. L., Smith, J. C., and Wilson, S. W. (2000) Silberblick/Wnt11 mediates convergent extension movements during zebrafish gastrulation. *Nature* **405**, 76–81.

10. Sokol, S. (2000) A role for Wnts in morphogenesis and tissue polarity. *Nat Cell Biol* **2**, E124–125.

11. Tada, M., and Smith, J. C. (2000) Xwnt11 is a target of Xenopus Brachyury: regulation of gastrulation movements via Dishevelled, but not through the canonical Wnt pathway. *Development* **127**, 2227–2238.

12. Oishi, I., Suzuki, H., Onishi, N., Takada, R., Kani, S., Ohkawara, B., Koshida, I., Suzuki, K., Yamada, G., Schwabe, G. C., Mundlos, S., Shibuya, H., Takada, S., and Minami, Y. (2003) The receptor tyrosine kinase Ror2 is involved in non-canonical Wnt5a/JNK signalling pathway. *Genes Cells* **8**, 645–654.

13. Qian, D., Jones, C., Rzadzinska, A., Mark, S., Zhang, X., Steel, K. P., Dai, X., and Chen, P. (2007) Wnt5a functions in planar cell polarity regulation in mice. *Dev Biol* **306**, 121–133.

14. Nomachi, A., Nishita, M., Inaba, D., Enomoto, M., Hamasaki, M., and Minami, Y. (2008) Receptor tyrosine kinase Ror2 mediates Wnt5a-induced polarized cell migration by activating c-Jun N-terminal kinase via actin-binding protein filamin A. *J Biol Chem* **283**, 27973–27981.

15. Boutros, M., Paricio, N., Strutt, D. I., and Mlodzik, M. (1998) Dishevelled activates JNK and discriminates between JNK pathways in planar polarity and wingless signaling. *Cell* **94**, 109–118.

16. Yamanaka, H., Moriguchi, T., Masuyama, N., Kusakabe, M., Hanafusa, H., Takada, R., Takada, S., and Nishida, E. (2002) JNK functions in the non-canonical Wnt pathway to regulate convergent extension movements in vertebrates. *EMBO Rep* **3**, 69–75.

17. Derijard, B., Hibi, M., Wu, I. H., Barrett, T., Su, B., Deng, T., Karin, M., and Davis, R. J. (1994) JNK1: a protein kinase stimulated by UV light and Ha-Ras that binds and phosphorylates the c-Jun activation domain. *Cell* **76**, 1025–1037.

18. Cheyette, B. N., Waxman, J. S., Miller, J. R., Takemaru, K., Sheldahl, L. C., Khlebtsova, N., Fox, E. P., Earnest, T., and Moon, R. T. (2002) Dapper, a Dishevelled-associated antagonist of beta-catenin and JNK signaling, is required for notochord formation. *Dev Cell* **2**, 449–461.

19. Pukrop, T., Klemm, F., Hagemann, T., Gradl, D., Schulz, M., Siemes, S., Trumper, L., and Binder, C. (2006) Wnt 5a signaling is critical for macrophage-induced invasion of breast cancer cell lines. *Proc Natl Acad Sci USA* **103**, 5454–5459.

20. Nishita, M., Itsukushima, S., Nomachi, A., Endo, M., Wang, Z., Inaba, D., Qiao, S., Takada, S., Kikuchi, A., and Minami, Y. (2010) Ror2/Frizzled complex mediates Wnt5a-induced AP-1 activation by regulating Dishevelled polymerization. *Mol Cell Biol.* **30**, 3610–3619.

21. Yamamoto, S., Nishimura, O., Misaki, K., Nishita, M., Minami, Y., Yonemura, S., Tarui, H., and Sasaki, H. (2008) Cthrc1 selectively activates the planar cell polarity pathway of Wnt signaling by stabilizing the Wnt-receptor complex. *Dev Cell* **15**, 23–36.

22. Kani, S., Oishi, I., Yamamoto, H., Yoda, A., Suzuki, H., Nomachi, A., Iozumi, K., Nishita, M., Kikuchi, A., Takumi, T., and Minami, Y. (2004) The receptor tyrosine kinase Ror2 associates with and is activated by casein kinase Iepsilon. *J Biol Chem* **279**, 50102–50109.

23. Scliwa, M., and Honer, B. (1993) Microtubules, centrosomes and intermediate filaments in directed cell movement. *Trends Cell Biol* **3**, 377–380.

24. Nobes, C. D., and Hall, A. (1999) Rho GTPases control polarity, protrusion, and adhesion during cell movement. *J Cell Biol* **144**, 1235–1244.

25. Lallemand, D., Ham, J., Garbay, S., Bakiri, L., Traincard, F., Jeannequin, O., Pfarr, C. M., and Yaniv, M. (1998) Stress-activated protein kinases are negatively regulated by cell density. *Embo J* **17**, 5615–5626.

26. Takada, R., Hijikata, H., Kondoh, H., and Takada, S. (2005) Analysis of combinatorial effects of Wnts and Frizzleds on beta-catenin/armadillo stabilization and Dishevelled phosphorylation. *Genes Cells* **10**, 919–928.

Chapter 17

Regulation of Focal Adhesion Dynamics by Wnt5a Signaling

Shinji Matsumoto and Akira Kikuchi

Abstract

Wnt5a is a representative ligand that activates the β-catenin-independent pathway of Wnt signaling in mammals. This pathway might be related to planar cell polarity signaling in *Drosophila*. Because reliable biochemical assays to measure Wnt5a pathway activity have not yet been established, we examined whether Wnt5a signaling stimulates focal adhesion turnover in migrating cells using live immunofluorescence imaging and immunocytochemical analysis. These assays demonstrated that the Wnt5a pathway cooperates with integrin signaling to regulate cell migration and adhesion through focal adhesion dynamics.

Key words: Wnt5a, Focal adhesion, Dvl, PCP, Migration, Integrin

1. Introduction

Most, if not all, cell types and tissues display several aspects of polarization. In addition to the ubiquitous epithelial cell polarity along the apical–basolateral axis, many epithelial tissues and organs are also polarized within the plane of the epithelium (1, 2). This is generally referred to as planar cell polarity (PCP; or tissue polarity). Genetic screens in *Drosophila* have identified core PCP factors such as *frizzled* (*fz*), *dishevelled* (*dvl*), *van gogh/strabismus*, and *prickle* (*pk*) (3, 4). These mutants display irregular hair orientation and abnormal rotation of ommatidia in the eye.

Fz and Dvl are also essential components of Wingless (Wg) signaling. Wg, Fz, and Dvl function as a cysteine-rich secreted ligand, a receptor, and an intracellular molecule, respectively, to regulate segment polarity and cell fate in *Drosophila* (5). Wg phenotypes exhibit low expression levels of engrailed in adjacent cells and disoriented cuticular structures in *Drosophila* embryos and an abnormal dorsal–ventral pattern of the wing disc. These phenotypes are different from PCP phenotypes. The Wg pathway

Kursad Turksen (ed.), *Planar Cell Polarity: Methods and Protocols*, Methods in Molecular Biology, vol. 839,
DOI 10.1007/978-1-61779-510-7_17, © Springer Science+Business Media, LLC 2012

stabilizes Armadillo (a β-catenin homolog), leading to expression of various kinds of target genes and is referred to as the canonical pathway or β-catenin pathway (6, 7). Wnt is a Wg homolog in vertebrates, and at least 19 Wnt members have been shown to be present in humans and mice. Wnt proteins exhibit unique expression patterns and have distinct functions during development, and some Wnts, including Wnt1, Wnt3, Wnt3a, and Wnt7a, activate the β-catenin pathway in mammals. It has been established that the Wg/Armadillo and Wnt/β-catenin pathways are well conserved between *Drosophila* and mammals.

In spite of extensive studies, it has not yet been clarified whether Wg itself is involved in PCP of *Drosophila* as an extracellular ligand. Fz and Dvl were proven to be involved in PCP genetically. It has been shown that DRho and DRock, both of which regulate cytoskeleton arrangement, act downstream of Dvl in PCP (8, 9), suggesting that another pathway (the PCP pathway or β-catenin-independent pathway) exists downstream of Dvl in addition to the Wg/β-catenin pathway.

Convergent extension is the intercalation and directed migration process driving mediolateral convergence and anteroposterior extension of the body axis during gastrulation of *Xenopus laevis* or zebrafish (3). Dvl has three conserved domains, the DIX, PDZ, and DEP domains (10). The DIX domain is necessary for *Drosophila* segment polarity and mammalian β-catenin stabilization, while the DEP domain of Dvl is required for *Drosophila* PCP and vertebrate convergent extension. Overexpression of some Wnts, including Wnt4, Wnt5a, and Wnt11, causes defective convergent extension movements without affecting cell fates in *Xenopus* embryos (11, 12). In addition, the zebrafish *silberblick* (*slb*) mutation, which has a defect in the morphogenesis of the prechordal plate, is an allele of *wnt11* (13). The *pipetail* (*ppt*) mutation, which is an allele of *wnt5*, has morphogenic defects in tail extension (14). Relatively mild *slb* mutant phenotypes can be rescued with the DEP domain of Dvl, suggesting that there are similarities between the PCP pathway in *Drosophila* and the Wnt5a/Wnt11 pathway in vertebrates. Therefore, in vertebrate cells it is considered that Wnt5a or Wnt11 signaling may regulate cell migration with polarity probably through the PCP pathway (β-catenin-independent pathway).

It has been reported that Wnt5a activates multiple signaling pathways to regulate cell migration, polarity, and cell-to-substrate attachment (15). So far we have found that knockdown of Wnt5a suppresses focal adhesion turnover and cell migration activity in various cells (16–19). Here, we describe methods to show focal adhesion turnover regulated by Wnt5a and its signaling pathway.

2. Materials

2.1. Cell Culture

1. Dulbecco's modified Eagle's medium (DMEM) (Nissui, Tokyo, Japan) supplemented with 10% fetal bovine serum (FBS) (Invitrogen, Carlsbad, CA, USA).

2. PenicillinG (add 0.325 g to 5 l of DMEM complete culture media) and streptomycin (add 0.5 g to 5 l of DMEM complete culture media) (Sigma-Aldrich, St. Louis, MO, USA).

3. Trypsin.

4. 35-mm culture dishes (BD Biosciences, San Jose, CA, USA).

5. Cell lines: HeLaS3 (human cervical cancer cell) cells (provided by K. Matsumoto, Nagoya University, Nagoya, Japan). Vero (monkey kidney epithelial cell) cells (provided by K. Kaibuchi, Nagoya University, Nagoya, Japan).

6. Cell culture incubator, 37°C, 5% CO_2.

2.2. Plasmid and siRNA Transfection

1. Transfection medium: Opti-MEM (Invitrogen).

2. Lipofectamin 2000 (Invitrogen).

3. Oligofectamin (Invitrogen).

4. The siRNA duplexes:
   Human Wnt5a-1 (sense), 5′-GUUCAGAUGUCAGAAGUAU-3′
   Human Wnt5a-2 (sense), 5′-GUGGAUAACACCUCUGUUU-3′
   Human Dvl1 (sense), 5′-GGAGGAGAUCUUUGAUGAC-3′
   Human Dvl2 (sense), 5′-GGAAGAAAUUUCAGAUGAC-3′
   Human Dvl3 (sense), 5′-GGAGGAGAUCUCGGAUGAC-3′
   Human Frizzled2 (Fz2) (sense), CGGUCUACAUGAUCAAAUA.

2.3. Live-Cell Imaging Analysis

1. 35-mm glass-based dishes (IWAKI, Tokyo, Japan).

2. Phenol-red free MEM (Invitrogen).

3. Type 1 collagen (Nutacon, Leimuiden, The Netherlands).

4. Olympus microscope IX81 (OLYMPUS, Tokyo, Japan).

2.4. Immunocyto-chemistry

1. Phosphate-buffered saline (PBS).

2. 4% (w/v) paraformaldehyde (in PBS).

3. 0.2% (w/v) Triton X-100 and 2 mg/ml BSA (in PBS).

4. 2 mg/ml BSA (in PBS).

5. Primary antibodies. Anti-paxillin (1:500 PBS) (BD Biosciences, Cat #610052). Anti-Wnt5a/b (1:100 PBS) (Cell Signaling Technology, Danvers, CA, USA, Cat #2530S). Anti-β1 integrin (1:100 PBS) (Millipore, Billenica, MA, USA, Cat #MAB2253Z). Anti-monoclonal FLAG (1:2000 PBS) (Sigma-Aldrich, Cat #F3165) or polyclonal FLAG (1:500 PBS) (Sigma-Aldrich, Cat #F7425) for FLAG-Fz2 staining.

6. Secondary Antibodies. Alexa Fluor 488 anti-mouse IgG (1:500 PBS) (Invitrogen, Cat #A11034). Alexa Fluor 546 anti-mouse IgG (1:500 PBS) (Invitrogen, Cat #A11030). Alexa Fluor 488 anti-rabbit IgG (1:500 PBS) (Invitrogen, Cat #A11029). Alexa Fluor 546 anti-rabbit IgG (1:500 PBS) (Invitrogen, Cat #A11035).

7. Alexa Fluor 546-phalloidin (1:500 PBS) (Invitrogen, Cat #A22283).

8. 18×18-mm glass cover slips (MATSUNAMI, Osaka, Japan).

9. 50% glycerol (in PBS).

10. Confocal microscope LSM510 (Carl-Zeiss, Jana, Germany).

2.5. Western Blot Analysis

1. NP-40 lysis buffer: 20 mM Tris–HCl pH 8.0, 1% NP40, 10% glycerol, and 137 mM NaCl. Add 10 µg/ml aprotinin, 1 mM PMSF, 10 µg/ml leupeptin, 5 mM NaF, 1 mM sodium *ortho*-vanadate, and 10 mM β-glycerophosphate to the lysis buffer prior to cell lysis.

2. 3× sodium dodecyl sulfate (SDS) loading buffer: 0.2 M Tris–HCl pH 6.8, 9.5% SDS, 16% glycerol, 0.015% bromophenol blue. Add 2-mercaptoethanol to the loading buffer at the final concentration of 6% prior to use.

3. 30% Acrylamide/bisacrylamide (Sigma-Aldrich) solution.

4. Ammonium persulfate (APS) (WAKO, Osaka, Japan).

5. TEMED (WAKO).

6. Dual color prestained ladder (Bio-Rad, Tokyo, Japan).

7. Running buffer: 25 mM Tris–HCL, 192 mM glycine, 0.1% SDS (w/v).

8. Transfer buffer: 25 mM Tris, 192 mM glycine, 20% methanol.

9. Nitrocellulose membrane (Whatman, Dassel, Germany).

10. Tris-buffered saline with Tween 20 (TBST).

11. 5% nonfat milk (in TBST).

12. Primary antibodies. Anti-FAK (1:1000 TBST) (BD Biosciences, Cat #610087). Anti-paxillin (1:1000 TBST) (BD Biosciences, Cat #610052). Anti-phospho-FAK (Tyr397) (1:1000 TBST) (Invitrogen, Cat #44-625G). Anti-phospho-paxillin (Tyr118) (1:1000 TBST) (Cell Signaling Technology, Cat #2541S).

13. Secondary antibodies. Horseradish peroxidase-conjugated anti-mouse IgG (1:5000 TBST) (GE Healthcare UK Ltd, Buckinghamshire, UK, Cat #NXA931). Horseradish peroxidase-conjugated anti-rabbit IgG (1:5000 TBST) (GE Healthcare UK Ltd, Cat #NA934).

14. HRP Substrate (GE Healthcare UK Ltd).

15. 1.5-ml tube (WATSON, Tokyo, Japan).

3. Methods

3.1. Cell Culture

HeLaS3 and Vero cells are cultured in DMEM supplemented with 10% FBS at 37°C with 5% CO_2.

3.2. Preparation of Cells for Transfection

3.2.1. siRNA Experiment

HeLaS3 cells (1×10^5 cells) are seeded onto 35-mm dish (BD Biosciences) and cultured in growth medium for 12 h. Then the medium is replaced with Opti-MEM (Invitrogen), and the cells are transfected with a mixture of siRNAs against Wnt5a-1 and -2, Dvl1, Dvl2, and Dvl3, or Fz2 at 40 nM each using Oligofectamine (Invitrogen). Serum was added into Opti-MEM at 4 h after transfection. The cells are used for experiments at 72 h after transfection.

3.2.2. Plasmid Transfection

When a plasmid is transfected to Vero cells, the medium is replaced with Opti-MEM (Invitrogen), and the cells are transfected with the indicated plasmids using Lipofectamine 2000 (Invitrogen). Then medium is replaced with DMEM containing 10% FBS 2–4 h after transfection. The cells are used for immunocytochemistry at 24 h after transfection.

3.3. Live-Cell Imaging Analysis of Focal Adhesions

It is well known that turnover of focal adhesions is required for a cell to spread and migrate (20). Spreading and migrating cells constantly assemble and disassemble their adhesions at the leading edge by the activation of small G proteins and FAK (21). Paxillin is a focal adhesion-associated adaptor molecule and a substrate for the Src–FAK complex (22).

Turnover of GFP-paxillin in HeLaS3 cells was analyzed by live fluorescence imaging (Fig. 1a). GFP-paxillin-containing adhesions disappeared as new adhesions were formed near the leading edge in wound-scratch-dependent migration. Disappearance and appearance of paxillin at the adhesion sites were reduced in Wnt5a knockdown cells. The results demonstrated that the average rates of disassembly and assembly of GFP-paxillin at adhesion sites were decreased in Wnt5a knockdown HeLaS3 cells (Fig. 1a).

Dvl is a critical component of Wnt signaling and mediates both the β-catenin-dependent and -independent pathways (10). In control cells, GFP-paxillin at the cell periphery turned over quickly within 45 min during spontaneous cell migration and exhibited multicolored signals in the merged image (Fig. 1b). In contrast, focal adhesions turned over poorly in Dvl knockdown cells and exhibited white color signals in the merged image (Fig. 1b).

3.3.1. Live Imaging of Focal Adhesions in Wound-Scratch-Dependent Migration

1. HeLaS3 cells (1×10^6 cells) expressing GFP-paxillin are plated onto 35-mm glass-based dish (IWAKI, Tokyo, Japan) coated with 10 μg/ml type 1 collagen (Nutacon, Leimuiden, The Netherlands) and cultured for 24 h.

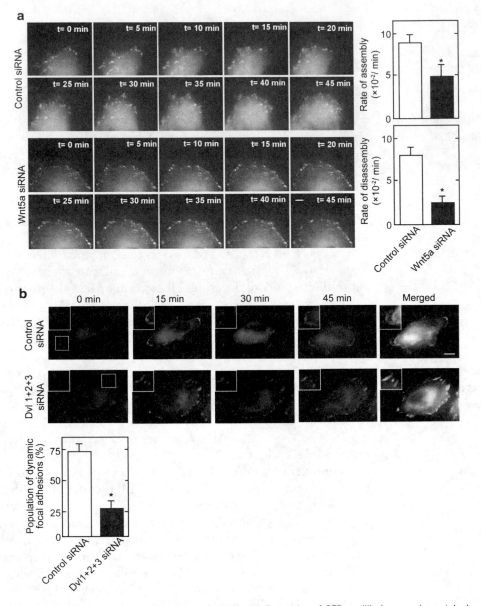

Fig. 1. Wnt5a signaling is involved in the dynamics of paxillin. (**a**) Dynamics of GFP-paxillin in wound-scratch-dependent migration of HeLaS3 cells transfected with control or Wnt5a siRNA were visualized by time-lapse fluorescence microscopy. Rate constants for disassembly and assembly of GFP-paxillin were calculated using MetaMorph software, and the results are shown in the *right-hand panels*. Scale bar, 5 μm. *$P < 0.05$. (**b**) Dynamics of GFP-paxillin in spontaneous migration of HeLaS3 cells transfected with control or Dvl siRNA were visualized by time-lapse fluorescence microscopy. The percentages of adhesions turning over within 45 min were calculated and the results are shown in the *bottom panel*. The region in the *white box* is shown enlarged. Scale bar, 10 μm. *$P < 0.01$. (**a**) was reproduced from Biochem. J. 2007 with permission from Portland Press Ltd, and (**b**) was reproduced from EMBO J. 2010 with permission from Nature Publishing Group.

2. The cell monolayer is then scratched manually using a plastic pipette tip.

3. Cells are washed with PBS, and the medium is replaced with phenol-red free MEM containing 10% FBS.

4. Four hours after wounding, images are captured using a fluorescence microscope (IX81).

5. The fluorescence intensity of individual adhesions from background-subtracted images is measured over time using MetaMorph software (Universal Imaging Corporation, Downingtown, PA, USA) (see Note 1). For rate constant measurements, the time length for disassembly (decreasing fluorescence intensity) of adhesions containing GFP-paxillin are plotted on separate semilogarithmic graphs representing fluorescence intensity ratios. Semilogarithmic plots of fluorescence intensity as a function of time are generated using the formula $\ln (I_0/I)$ for disassembly, where I_0 is the initial fluorescence intensity and I is the fluorescence intensity at various time points. The slope of the linear regression trend line fitted to the semilogarithmic plots is then calculated to determine apparent rate constants of disassembly. For each rate constant, measurements are made for at least ten individual adhesions in five separate cells.

3.3.2. Live Imaging of Focal Adhesions in Spontaneous Migration

1. HeLaS3 cells expressing GFP-paxillin are sparsely plated onto collagen-coated 35-mm glass-based dishes and cultured in phenol-red free MEM containing 10% FBS.

2. Time lapse images are collected from 2 h after cell plating using fluorescence microscope (IX81) and MetaMorph software (see Note 2).

3. The images from time points at 0, 15, 30, and 45 min are represented in red, green, blue, and white, respectively, and are superimposed. White spots are counted as nondynamic adhesions, and spots containing different colors are counted as dynamic adhesions. The percentage of adhesions turning over within 45 min is calculated for ten different focal adhesions per cell (counted cells: 20 per siRNA).

3.4. Immunocyto-chemistry of Intracellular Molecules

Because it has been suggested that the formation of large focal adhesions reflects the reduced turnover of adhesions (23), the effects of Wnt5a and Dvl on the size of focal adhesions were examined by using immunocytochemical staining of endogenous paxillin. Knockdown of Wnt5a and Dvl induced the enlargement of focal adhesions in HeLaS3 cells (Fig. 2a). The addition of Wnt5a to cells increased the number of focal adhesions, reflecting its involvement in increasing the rates of focal adhesion turnover (Fig. 2b). Furthermore, knockdown of FZ2, a Wnt5a receptor,

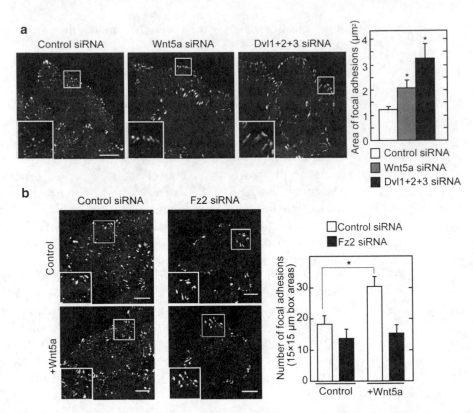

Fig. 2. Wnt5a signaling stimulates the dynamics of focal adhesion. (**a**) The sizes of focal adhesions in HeLaS3 cells transfected with control, Wnt5a, or Dvl siRNA were analyzed using immunocytochemical staining of endogenous paxillin. The individual areas of paxillin staining were quantified using LSM Image browser software (Carl Zeiss), and the results are shown in the *right-hand panel*. (**b**) HeLaS3 cells transfected with control or Fz2 siRNA were stimulated with Wnt5a conditioned medium for 60 min. The numbers of focal adhesions were counted using immunocytochemical staining of endogenous paxillin. The numbers of paxillin stained region in three different areas (15 × 15 μm) at the cell periphery (*white box*) were counted using LSM Image browser software, and the results are shown in the *right-hand panel*. Scale bar, 10 μm. *$P < 0.01$. This figure was reproduced from EMBO J. 2010 with permission from Nature Publishing Group.

inhibited the effect of the Wnt5a-dependent increase in focal adhesion numbers (Fig. 2b). Taken together, these results showed that Wnt5a and Dvl signaling stimulates focal adhesion dynamics and that the Wnt5a signal requires Fz2 in HeLaS3 cells.

1. HeLaS3 cells grown on glass coverslips are fixed for 10 min at room temperature in PBS containing 4% (w/v) paraformaldehyde.

2. The cells are permeabilized in PBS containing 0.2% (w/v) Triton X-100 and 2 mg/ml BSA for 10 min at room temperature.

3. The cells are blocked in PBS containing 2 mg/ml BSA for 30 min at room temperature.

4. The cells are incubated with an antibody against paxillin for 1 h at room temperature.

5. The samples are incubated with Alexa Fluor 488-conjugated antibodies against mouse IgG and with Alexa Fluor 546-phalloidin for 1 h at room temperature.

6. The samples are viewed using a confocal microscope (LSM510).

3.5. Immunocyto-chemistry of Cell Surface Molecules

To visualize the localization of Wnt5a, Wnt5a was expressed in directional migrating Vero cells. When Wnt5a was expressed alone, it could not be detected at the cell surface (data not shown). When Wnt5a and FLAG-Fz2 were expressed simultaneously, both proteins colocalized to the cell front (leading edge) and also to the boundary of cell-to-cell adhesions (Fig. 3a, top panels). To show the colocalization of Wnt5a and Fz2 more clearly, their expression at the dorsal cell surface was examined. FLAG-Fz2 was indeed colocalized with Wnt5a as punctate structures (Fig. 3a, bottom panels). These results suggested that Wnt5a/Fz2 signaling has a potential role at the leading edge in motile cells.

A small size of the paxillin-positive staining was observed at the leading edge of polarized Vero cells, while a large area of paxillin staining was detected in other regions (Fig. 3b). This reflected the rapid turnover of focal complexes containing paxillin at the leading edge of motile cells and the slow turnover of focal adhesions in other regions (23).

Integrins are essential molecules to regulate cell-substrate adhesion and migration (24). Immunocytochemical analysis showed that endogenous cell-surface $\beta1$ integrin accumulated at the leading edge where cell-surface FLAG-Fz2 also accumulated in migrating Vero cells (Fig. 3c, top panels). When the localization of $\beta1$ integrin and FLAG-Fz2 at the ventral (substrate-facing) cell surface was examined using a confocal microscope, $\beta1$ integrin puncta localized adjacent to the puncta of FLAG-Fz2 (Fig. 3c, bottom panels). These results raised the possibility that there is a connection between Wnt5a/Fz2 and integrin signaling.

3.5.1. Localization of Wnt5a and Fz2

1. Vero cells grown on glass coverslips are transfected with Wnt5a and FLAG-Fz2.

2. The cells are incubated in fresh growth medium containing an anti-Wnt5a and anti-monclonal FLAG antibodies at 37°C with 5% CO_2 for 1 h before fixation (see Note 3).

3. The cells are fixed for 10 min at room temperature in PBS containing 4% (w/v) paraformaldehyde.

4. The cells are further incubated with Alexa Fluor 488- or 546-conjugated antibodies against rabbit and mouse IgG for 1 h at room temperature.

5. The samples are viewed using a confocal microscope (LSM510).

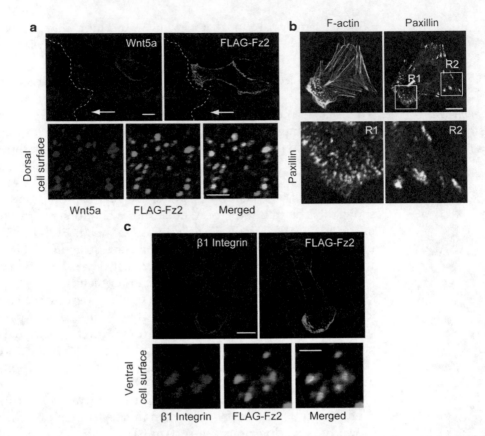

Fig. 3. Fz2 colocalizes with Wnt5a and β1 integrin at the leading edge in Vero cells. (**a**) Vero cells expressing FLAG-Fz2 and Wnt5a were wounded. After 4 h, the cells were incubated with anti-FLAG and anti-Wnt5a antibodies for 1 h before fixation. The cells were fixed and probed with Alexa 546-conjugated anti-rabbit IgG and Alexa 488-conjugated anti-mouse IgG antibodies. *White arrows* indicate the migration direction. *Dashed lines* indicate the front line of scratched cells. Colocalization of punctuate structures of FLAG-Fz2 and Wnt5a were observed at the dorsal cell surface (*bottom panels*). (**b**) Vero cells were stained with anti-paxillin antibody and phalloidin. The region in the *white boxes* (R1 and R2) are shown enlarged in *bottom panels*. (**c**) Vero cells expressing FLAG-Fz2 were fixed and stained with anti-β1 integrin and anti-FLAG polyclonal antibodies without permeabilization (*top panels*). After the removal of free antibody, cells were probed with Alexa 488-conjugated anti-rabbit IgG and Alexa 546-conjugated anti-mouse IgG antibodies. Close localization of punctuate structure of FLAG-Fz2 and β1 integrin was observed at the ventral cell surface (*bottom panels*). Scale bars in top panels, 10 μm; bottom panels, 2 μm. This figure was reproduced from EMBO J. 2010 with permission from Nature Publishing Group.

3.5.2. Localization of Fz2 and β1 Integrin

1. Vero cells grown on glass coverslips are transfected with FLAG-Fz2.

2. The cells are fixed for 10 min at room temperature in PBS containing 4% (w/v) paraformaldehyde.

3. The cells are incubated with anti-β1 integrin and anti-polyclonal FLAG antibodies for 1 h at room temperature without permeabilization (see Note 3).

4. The cells are further incubated with Alexa Fluor 488- or 546-conjugated antibodies against rabbit and mouse IgG for 1 h at room temperature.

5. The samples are viewed using a confocal microscope (LSM510).

3.6. Western Blot Analysis for Adhesion-Dependent Activation of Focal Adhesions

When cells attach to the substrate, several tyrosine residues become phosphorylated upon FAK activation (25). FAK is activated via autophosphorylation at Tyr397, which is initiated by integrin engagement with its ligand (26), and paxillin is phosphorylated by FAK at Tyr118 (27). The cell adhesion-dependent activation of FAK and phosphorylation of paxillin were reduced in Wnt5a and Dvl knockdown HeLaS3 cells (Fig. 4).

1. HeLaS3 cells (1×10^6 cells) transfected with control, Wnt5a, or Dvl siRNA are suspended in serum-free medium (DMEM).

2. The suspension is transferred into a 1.5-ml tube (WATSON, Tokyo, Japan) and incubated for 1 h at 37°C.

3. They are further kept in suspension or plated onto a collagen-coated 35-mm dish (BD Biosciences) for 1 h at 37°C.

4. Whole cells in the dish are lysed using NP-40 lysis buffer.

5. The lysates are mixed with 3× SDS loading buffer.

6. Protein samples are separated by SDS-polyacrylamide gel electrophoresis and transferred onto nitrocellulose membranes.

7. The membranes are blocked in TBS containing 0.5% Tween 20 and 5% nonfat milk for 1 h at room temperature.

8. The membranes are incubated with anti-FAK, anti-paxillin, anti-phospho-FAK (Tyr397), or anti-phospho-paxillin (Tyr118) antibody for 1 h at room temperature.

9. The membranes are further incubated with horseradish peroxidase-conjugated anti-mouse IgG or anti-rabbit IgG for 1 h at room temperature.

10. Detection of antibody reactivity was performed using Western HRP Substrate.

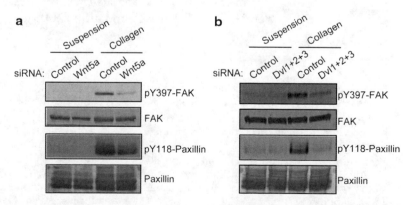

Fig. 4. Knockdown of Wnt5a and Dvl suppresses adhesion-dependent activation of FAK and phosphorylation of paxillin. After HeLaS3 cells transfected with control, Wnt5a (**a**), or Dvl (**b**) siRNA were suspended in serum-free medium for 1 h, they were kept in suspension or plated onto collagen-coated dishes for 1 h. Lysates were probed with anti-pY397-FAK and anti-pY118-paxillin antibodies. FAK and paxillin were used as loading controls. This figure was reproduced from EMBO J. 2010 with permission from Nature Publishing Group.

4. Notes

1. To assess wound-scratch-dependent turnover of GFP-paxillin, focal adhesions located at the free edges of the cell monolayer were observed.

2. When turnover of GFP-paxillin or the area of individual focal adhesions in the spontaneous migration of HeLaS3 cells was examined, it was necessary to observe cells that were morphologically similar, because HeLaS3 cells are difficult to polarize. In our experiment, focal adhesions located at the cell edge were examined.

3. Although Wnt5a is a secreted protein, it was difficult to observe the localization of Wnt5a at the cell surface. The only way that we succeeded was to express Wnt5a and FLAG-Fz2 simultaneously (Fig. 3a). In this experiment, cells were initially incubated with anti-Wnt5a and anti-FLAG antibodies in DMEM containing 10% FBS. After the removal of free anti-Wnt5a and anti-FLAG antibodies, the cells were fixed and then incubated with the secondary antibodies.

Acknowledgments

We thank Nature Publishing Group and Portland Press Ltd for permitting the use of original figures. Financial support was provided by Grants-in-Aid for Scientific Research and for Scientific Research on Priority Areas from the Ministry of Education, Science, and Culture of Japan (2008, 2009, 2010), and by Research Grants from the Princess Takamatsu Cancer Research Fund (08-24005) and Takeda Science Foundation (2009).

References

1. Adler, P. N. (2002) Planar signaling and morphogenesis in *Drosophila*. *Dev Cell* **2**, 525–35.

2. Axelrod, J. D. (2009) Progress and challenges in understanding planar cell polarity signaling. *Semin Cell Dev Biol* **20**, 964–71.

3. Veeman, M. T., Axelrod, J. D., and Moon, R. T. (2003) A second canon. Functions and mechanisms of β-catenin-independent Wnt signaling. *Dev Cell* **5**, 367–77.

4. Wu, J., and Mlodzik, M. (2009) A quest for the mechanism regulating global planar cell polarity of tissues. *Trends Cell Biol* **19**, 295–305.

5. Cadigan, K. M., and Nusse, R. (1997) Wnt signaling: a common theme in animal development. *Genes Dev* **11**, 3286–305.

6. Logan, C. Y., and Nusse, R. (2004) The Wnt signaling pathway in development and disease. *Annu Rev Cell Dev Biol* **20**, 781–810.

7. Kikuchi, A., Yamamoto, H., and Sato, A. (2009) Selective activation mechanisms of Wnt signaling pathways. *Trends Cell Biol* **19**, 119–29.

8. Strutt, D. I., Weber, U., and Mlodzik, M. (1997) The role of RhoA in tissue polarity and Frizzled signalling. *Nature* **387**, 292–5.

9. Winter, C. G., Wang, B., Ballew, A., et al. (2001) *Drosophila* Rho-associated kinase (Drok) links Frizzled-mediated planar cell polarity signaling to the actin cytoskeleton. *Cell* **105**, 81–91.

10. Wharton, K. A. J. (2003) Runnin' with the Dvl: proteins that associate with Dsh/Dvl and their significance to Wnt signal transduction. *Dev Biol* **253**, 1–17.

11. Du, S. J., Purcell, S. M., Christian, J. L., McGrew, L. L., and Moon, R. T. (1995) Identification of distinct classes and functional domains of Wnts through expression of wild-type and chimeric proteins in Xenopus embryos. *Mol Cell Biol* **15**, 2625–34.

12. Moon, R. T., Campbell, R. M., Christian, J. L., McGrew, L. L., Shih, J., and Fraser, S. (1993) Xwnt-5A: a maternal Wnt that affects morphogenetic movements after overexpression in embryos of *Xenopus laevis*. *Development* **119**, 97–111.

13. Heisenberg, C. P., Tada, M., Rauch, G. J., et al. (2000) Silberblick/Wnt11 mediates convergent extension movements during zebrafish gastrulation. *Nature* **405**, 76–81.

14. Rauch, G. J., Hammerschmidt, M., Blader, P., et al. (1997) Wnt5 is required for tail formation in the zebrafish embryo. *Cold Spring Harb Symp Quant Biol* **62**, 227–34.

15. Kikuchi, A., and Yamamoto, H. (2008) Tumor formation due to abnormalities in the β-catenin-independent pathway of Wnt signaling. *Cancer Sci* **99**, 202–8.

16. Kurayoshi, M., Oue, N., Yamamoto, H., et al. (2006) Expression of Wnt-5a is correlated with aggressiveness of gastric cancer by stimulating cell migration and invasion. *Cancer Res* **66**, 10439–48.

17. Kurayoshi, M., Yamamoto, H., Izumi, S., and Kikuchi, A. (2007) Post-translational palmitoylation and glycosylation of Wnt-5a are necessary for its signalling. *Biochem J* **402**, 515–23.

18. Yamamoto, H., Kitadai, Y., Oue, N., Ohdan, H., Yasui, W., and Kikuchi, A. (2009) Laminin γ2 mediates Wnt5a-induced invasion of gastric cancer cells. *Gastroenterology* **137**, 242–52, 52 e1–6.

19. Matsumoto, S., Fumoto, K., Okamoto, T., Kaibuchi, K., and Kikuchi, A. (2010) Binding of APC and dishevelled mediates Wnt5a-regulated focal adhesion dynamics in migrating cells. *Embo J* **29**, 1192–204.

20. Small, J. V., and Kaverina, I. (2003) Microtubules meet substrate adhesions to arrange cell polarity. *Curr Opin Cell Biol* **15**, 40–7.

21. Mitra, S. K., Hanson, D. A., and Schlaepfer, D. D. (2005) Focal adhesion kinase: in command and control of cell motility. *Nat Rev Mol Cell Biol* **6**, 56–68.

22. Carragher, N. O., and Frame, M. C. (2004) Focal adhesion and actin dynamics: a place where kinases and proteases meet to promote invasion. *Trends Cell Biol* **14**, 241–9.

23. Ridley, A. J., Schwartz, M. A., Burridge, K., et al. (2003) Cell migration: integrating signals from front to back. *Science* **302**, 1704–9.

24. Legate, K. R., Wickstrom, S. A., and Fassler, R. (2009) Genetic and cell biological analysis of integrin outside-in signaling. *Genes Dev* **23**, 397–418.

25. Wozniak, M. A., Modzelewska, K., Kwong, L., and Keely, P. J. (2004) Focal adhesion regulation of cell behavior. *Biochim Biophys Acta* **1692**, 103–19.

26. Schaller, M. D., Hildebrand, J. D., Shannon, J. D., Fox, J. W., Vines, R. R., and Parsons, J. T. (1994) Autophosphorylation of the focal adhesion kinase, pp125FAK, directs SH2-dependent binding of pp60src. *Mol Cell Biol* **14**, 1680–8.

27. Bellis, S. L., Perrotta, J. A., Curtis, M. S., and Turner, C. E. (1997) Adhesion of fibroblasts to fibronectin stimulates both serine and tyrosine phosphorylation of paxillin. *Biochem J* **325** **(Pt 2)**, 375–81.

Chapter 18

The Embryonic Mouse Gut Tube as a Model for Analysis of Epithelial Polarity

Makoto Matsuyama and Akihiko Shimono

Abstract

Recent accumulating data indicate links between planer cell polarity (PCP) and apicobasal (AB) polarity in epithelial cells. PCP regulatory genes have been shown to be involved in the establishment of AB polarity in addition to regulating PCP. We have shown that the gut tube of the mouse embryo is a unique model system for the analysis of epithelial polarities, e.g., oriented cell division and AB polarity, with respect to the PCP pathway. The regulation of epithelial polarity by the PCP pathway might play an essential role in organ morphogenesis relating to physiological function.

Key words: Apicobasal (AB) polarity, Oriented cell division, Gut tube, Confocal microscope, Immunofluorescent staining

1. Introduction

Neural tube defects, such as craniorachischisis and spina bifida, and defects in convergent extension of the gastrula embryo are widely observed phenotypes in animals with mutations in PCP pathway component genes (1, 2). Additionally, the orientation of sensory hair cells of the cochlea has frequently been utilized as a model system for studying PCP regulation in mammals (3). Deviations in the orientation of the stereocilia bundles of hair cells, or incidence of these other phenotypes in single or compound mutant animals are hallmarks which confirm that a gene is involved in PCP regulation. In addition to the broader scale events of tissue morphogenesis, analysis at the cellular level can provide great insight into the molecular mechanisms underlying PCP signaling. Although the relationship between the higher events of tissue morphogenesis and cellular events is still unclear, recent evidence derived from

Kursad Turksen (ed.), *Planar Cell Polarity: Methods and Protocols*, Methods in Molecular Biology, vol. 839,
DOI 10.1007/978-1-61779-510-7_18, © Springer Science+Business Media, LLC 2012

mouse genetics indicates that PCP core components are involved in oriented cell division and apicobasal (AB) polarity (4). Thus, the PCP pathway regulates both PCP and AB polarity.

While cell polarity regulation via the PCP pathway might be crucial in organ formation (3, 5, 6), aberrant regulation can also result in human disease—defects in oriented cell division are involved in polycystic kidney disease (3) and disrupted cell polarity contributes to invasion and metastasis of cancer cells (7). The gut tube is a typical epithelial tissue maintaining its epithelial polarity from embryonic development through adult tissue homeostasis. We have shown that the embryonic gut tube is a unique system for analysis of epithelial polarity regulated via the PCP pathway (8). Confocal images can be obtained from whole-mount preparations of stomach epithelium immunostained to visualize mitotic spindles and chromosomes with respect to oriented cell division. Several antibodies are commercially available to observe AB polarity as well as subcellular distribution of PCP core components in gut epithelium.

2. Materials

2.1. Tissue Preparation

1. Phosphate-buffered saline (PBS).
2. Dissection solution: 10% (v/v) fetal calf serum (FCS) in PBS.
3. Fixative: 4% (w/v) paraformaldehyde in PBS (PFA/PBS) (see Note 1).
4. Methanol/PBS series: 25, 50, and 75% methanol in PBS.
5. 100% Methanol for storage of tissue samples.

2.2. Tissue Sections

1. HEPES-Buffered Saline Solution (HBSS): 10 mM HEPES, 1 mM $CaCl_2$, 1 mM $MgSO_4$, 137 mM NaCl, 5.37 mM KCl, 5.56 mM Glucose, 0.335 mM $Na_2HPO_4 \cdot 12H_2O$. Add phenol red at 1 mg/l and adjust pH to 7.5 using 1N NaOH (see Note 2).
2. Sucrose/HBSS solutions: 10, 15, 25% (w/v) sucrose in HBSS.
3. O.C.T. compound (Tissue-Tek).
4. Cryomold (Tissue-Tek).
5. Liquid N_2 or a dry ice block.
6. MAS-coated slide glass (Matsunami).

2.3. Immunofluorescent Staining

1. TBS: 140 mM NaCl, 2.7 mM KCl, 25 mM Tris–HCl, pH 7.5.
2. TBSTx: 1% (v/v) TritonX-100 in TBS.
3. TBST: 0.1% (v/v) Tween-20 in TBS.
4. Dako pen (Dako).

5. Blocking solution: for whole-mount, 5% (w/v) nonfat dried skim milk (Difco) in TBSTx; for sections, 3% (w/v) bovine serum albumin (BSA, Sigma) in TBST.

6. Primary antibodies: for whole-mount immunostaining, anti-acetylated-tubulin (mouse monoclonal antibody, 1:4,000, Sigma), anti-α-tubulin (mouse monoclonal antibody, 1:5,000, Sigma), anti-γ-tubulin (mouse monoclonal antibody, 1:5,000, Sigma), anti-β1-integrin (rat monoclonal antibody: clone MB1.2, 1:1,000, Chemicon); for sections, anti-phospho-aPKC (from rabbit, 1:100, Cell Signaling), anti-E-cadherin (rat monoclonal anti-Uvomorullin antibody, 1:1,000, Sigma), anti-Frizzled3 (from rabbit, 1:50, MBL), anti-Disheveled-2 (from goat, sc-7399, 1:50, Santa Cruz), anti-β1-integrin (rat monoclonal antibody: clone MB1.2, 1:100, Chemicon).

7. Secondary antibodies: Alexa Flour 488 or 594 conjugated donkey anti-rabbit IgG (H+L), Alexa Flour 488 or 594 conjugated goat anti-mouse IgG (H+L), Alexa Flour 594 conjugated donkey anti-goat or -rat IgG (H+L) (Invitrogen).

8. 4′,6-Diamidino-2-phenylindole (DAPI, Sigma): Dissolve DAPI in H_2O at 2 mg/ml and store at –20°C. To use, dilute the stock solution to working concentration.

9. Mounting medium: e.g., Antifade (Invitrogen) or Crystal Mount (Biomeda).

2.4. Imaging

1. Mounting slide (Fig. 1c; made-to-order).

2. Nail varnish.

3. Confocal microscope.

4. Software: Photoshop (Adobe), Image Browser (Zeiss; see Note 3).

3. Methods

The methods described in this section are immunostaining procedures using antibodies on whole mount or sectioned samples. The target proteins are expressed in the epithelial tissues or cells with unique localization patterns. It is possible that proteins might degrade during the time-consuming dissection process or the subcellular distribution of proteins might destabilize under mechanical forces. Therefore, we strongly recommend that tissue dissection procedures before fixation should be performed carefully; the dissection should be done in cold dissection solution, and the gastrointestinal tract should remain as intact as possible. Once samples are fixed and processed appropriately, they can be stored for several months at –20 or –80°C. Most of the antibodies

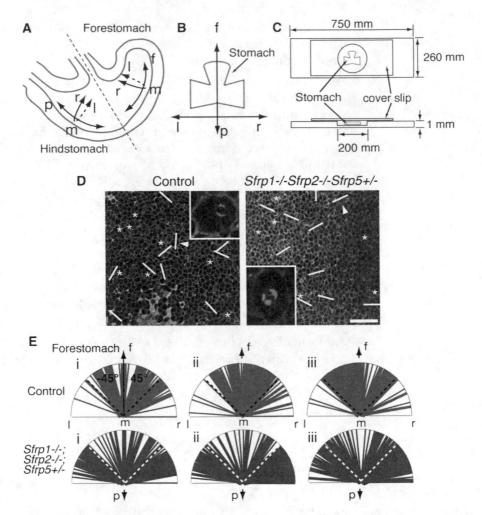

Fig. 1. Orientation of cell division in the epithelium of the embryonic stomach. (**a**) In this schematic diagram of an E13.5 stomach, *arrows* corresponding to f, p, r, and l depict the direction of the fundus, pylorus, right, and left, respectively. (**b**) A "Flattened" stomach. *Arrows* corresponding to f, p, r, and l are the same as in (**a**). (**c**) A mounting slide for whole-mounting embryonic stomach tissue. *Top figure* is a top-view, and *bottom* is a side-view of the mounting slide. The whole-mount stomach sample is placed between the bottom of a well and a cover slip. (**d**) Cell division orientation was visualized with anti-β1-integrin, anti-acetylated α-tubulin, and DAPI staining in the epithelium of the fore-stomach of control and *Sfrps* deficient embryos. The top of the confocal image is oriented to the direction of the fundus. The *inset* shows a higher magnification of the mitotic cells indicated by an *arrowhead*. The cell division axis is indicated by a *bar* whereas the vertical axis is denoted by an *asterisk*. Scale bar: 40 μm. (**e**) Schematic diagrams depicting the angles of cell division orientation as measured in three fore-stomachs (i, ii, and iii) from wild type and *Sfrps* deficient embryos, respectively. The orientation of cell division converges within ±45° in the control fore-stomach; however, it diverges in the *Sfrps* deficient fore-stomach (**a**, **d**, and **e–i** are reproduced from ref. 8 under Creative Commons Attribution License).

described here are commercially available. It is important to note that the specific fixation and blocking conditions depend on the primary antibody, thus if antibodies different from those listed in this protocol are to be used, these conditions should be taken into account.

3.1. Analyses of Oriented Cell Division in Stomach Epithelium

3.1.1. Preparation of Embryonic Mouse Stomach

1. Euthanize a pregnant mouse at embryonic day (E) 12.5–13.5 and collect the embryos in dissection solution keeping the yolk sac and placenta attached. Transfer an embryo, with extra-embryonic tissues attached, into fresh cold dissection solution. Remove the extra-embryonic tissues; subsequently, separate the head and limbs. Open the ventral surface ectoderm and remove the heart and liver with forceps while ensuring that the gut tube is left as intact as possible. Transfer the tissue containing the gut tube into 4% PFA/PBS at 4°C immediately after dissection. Keep the remaining embryos in cold dissection solution and process them one by one. Use at least a tenfold volume of fixative to the volume of the tissue(s).

2. Fix the samples in 4% PFA/PBS overnight at 4°C. After fixation, carefully dissect out the stomach which is discernible between the esophagus and duodenum. Transfer the tissues into ice-cold PBS using a transfer pipette. Avoid damaging the stomach.

3. Wash the tissues twice in PBS at room temperature (RT).

4. Transfer them into an ascending methanol/PBS series (25, 50, and 75%, respectively) for 5 min in each solution at RT, then in 100% methanol, twice.

5. Store the samples at –20°C for up to 6 months.

3.1.2. Whole-Mount Immunofluorescent Staining

1. Rehydrate the stomach by transferring it into a descending methanol/PBS series (75, 50, 25%) and finally in PBS for 5 min in each solution at RT.

2. Wash twice in TBSTx for 5 min at RT then transfer the stomach, together with TBSTx, using a transfer pipette into a 1.5-ml tube for subsequent antibody staining.

3. Replace the TBSTx with blocking solution. Incubate the sample in the solution for 60 min at RT while gently rocking.

4. Apply the primary antibody mixture (e.g., anti-β1-integrin and anti-acetylated-α-tubulin; see Note 4) diluted with the blocking solution overnight at 4°C with rocking.

5. The next day, wash the stomach three times with TBSTx for 5 min at RT.

6. Wash three times in TBSTx for 30 min at RT with rocking.

7. Wash three times in TBSTx for 60 min at RT with rocking.

8. Apply the secondary antibody mixture diluted (1:1,000) with the blocking solution overnight at 4°C with rocking. Keep the samples in the dark hereafter as much as possible.

9. The next day, wash the stomach three times in TBSTx for 5 min at RT.

10. Wash three times with TBSTx containing 0.02 μg/ml DAPI for 30 min at RT with rocking.

11. Wash three times with TBSTx for 60 min at RT with rocking.

12. Divide the stomach into the fore-stomach and hind-stomach under a dissection microscope. Cut the tissue from the esophageal side leaving the greater curvature intact. Carefully open the stomach so it lays flat, being mindful not to lose the orientation from the fundus to the pylorus (see Fig. 1b).

13. Place the "flattened" stomach into a well of a mounting slide (made-to-order) being careful not to lose the orientation (see Fig. 1c). Apply a few drops of TBSTx or mounting medium (e.g., Crystal Mount) to the slide and place a cover slip over the well (see Fig. 1c).

14. Affix the cover slip to the slide by sealing the edges with nail varnish.

3.1.3. Imaging to Measure the Orientation of Cell Division

1. Scan and obtain images of the epithelium of the greater curvature of the stomach using a confocal microscope. The direction from bottom to top of the confocal image should be adjusted to correspond to the direction from fundus to pylorus. An example of the results produced is shown in Fig. 1d.

2. Use software to analyze the images; in this case Image Browser software can be used (see Note 3). Open the data using Image Browser and select the "Overlay" panel. Draw a straight line to the data along the direction of cell division with respect to orientation along the axis from fundus to pylorus, and check the "Measure" panel to measure the angle of mitotic orientation. As shown in Fig. 1e, plot the angles of more than 100 mitotic epithelial cells from at least three different stomachs. The data should be statistically analyzed.

3. Alternatively, the image can be analyzed manually if the software is not available. Print the image on copy paper using Photoshop software, and find the divided cells in the epithelium of the stomach. Mark the direction of cell division on the paper. Measure the angle of the mitotic direction with respect to orientation along the axis from fundus to pylorus.

3.2. Analyses of Apicobasal Polarity in Gut Epithelium

3.2.1. Preparation of Cryo-sections of Embryonic Gut

1. Euthanize a pregnant mouse and dissect the embryos as described in Subheading 3.1.1, step 1.

2. Fix the samples in 4% PFA/PBS overnight at 4°C. After fixation, carefully separate the gastrointestinal tract from the other tissues and transfer it into ice-cold PBS using a transfer pipette with an opening cut to the appropriate size.

3. Wash the tissue twice by transferring it into clean PBS, then transfer into 10% sucrose/HBSS for 1 h or overnight at 4°C with rocking in order to suffuse the tissue with the sucrose solution.

4. Transfer the sample into 15% sucrose/HBSS and incubate with rocking for one to several hours at 4°C. Finally, transfer the tissue into 25% sucrose/HBSS and incubate for at least 1 h or overnight at 4°C with rocking.

5. Put the sample into O.C.T. compound in a small dish using a transfer pipette. Remove excess solution and transfer the sample in O.C.T. compound into an embedding mold (Cryomold) filled with O.C.T. Freeze sample on a dry ice block.

6. Store the frozen sample at –80°C until sectioning.

7. Make 5-μm thick sections using a cryostat and collect them on glass slides. Dry the sections at RT until the O.C.T. compound has dried completely. The slides can be stored at –80°C for a few months.

3.2.2. Immunofluorescent Staining of Sections

1. Air dry slide if condensation forms when removed from the freezer.

2. Label the slide with a pencil, noting date and primary antibody to be used. Draw a line of hydrophobic material around specimens using a Dako pen in order to retain the small volumes of antibody solutions on the slide.

3. Wash the slide in TBST for 5 min at RT, gently mixing the solution to remove the embedding medium from around the section.

4. Remove excess TBST from the slide and place slide in a staining tray. Apply 500 μl of blocking solution to a slide, and incubate for 30 min at RT.

5. Remove and discard excess blocking solution and apply primary antibody (e.g., anti-phospho-aPKC antibody and anti-β1-integrin in Fig. 2a, b, anti-Frizzled3 antibody and anti-Disheveled-2 antibody in Fig. 2c, d) diluted with blocking solution to the slide. Incubate sections with the antibody overnight at 4°C.

6. The next day, discard the primary antibody solution. Rinse the slide three times in TBST for 5 min at RT, gently mixing the solution.

7. Apply secondary antibody diluted (1:200) with blocking solution for 1 h at RT. Keep the slide in the dark hereafter as much as possible.

8. Rinse and simultaneously stain the sections with DAPI in TBST containing 0.02 μg/ml DAPI for 5 min at RT, gently mixing the solution. Subsequently, wash the slide twice in TBST for 5 min at RT.

9. Apply a few drops of mounting medium (e.g., Antifade) to a slide and place a cover slip over it.

10. Remove any excess mounting medium using small pieces of paper towel or filter paper. (It may be necessary to dry the

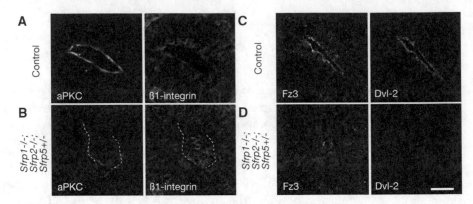

Fig. 2. AB polarity protein localization in the epithelium of the small intestine. (**a**, **b**) aPKC and β1-integrin subcellular localization in the epithelium of the small intestine derived from wild type (**a**) and *Sfrps* deficient embryos (**b**). (**c**, **d**) Frizzled3 (Fz3) and Disheveled-2 (Dvl-2) subcellular localization in the epithelium of the small intestine derived from wild type (**c**) and *Sfrps* deficient embryos (**d**). Scale bar: 25 μm.

mounting medium according to the manufacturer's manual; store the slide in the dark overnight at RT.)

11. Affix the cover slip to the slide by sealing the edges with nail varnish.

12. Observe staining under a florescent microscope after the nail varnish has dried. An example of the results produced is shown in Fig. 2. It is possible to store the stained sample in a dark box at 4°C for a few weeks.

4. Notes

1. It is important to use freshly made 4% PFA/PBS. Use the fixative within 12 h of preparation.

2. The addition of phenol red to the HBSS helps in identifying the sample in a frozen block when sectioning.

3. Image Browser software is freeware (http://www.zeiss.com).

4. Anti-β1-integrin antibody, anti-acetylated-α-tubulin antibody, and DAPI can be used to visualize cell membranes, two pairs of the mitotic spindle, and chromosomes, respectively (Fig. 1d).

Acknowledgments

This work was supported by the Singapore National Research Foundation and the Ministry of Education under the Research Center of Excellence Program.

References

1. Greene, N. D., Stanier, P., and Copp, A. J. (2009) Genetics of human neural tube defects. *Hum Mol Genet.* **18**, R113–129.

2. Seifert, J. R., and Mlodzik, M. (2007) Frizzled/PCP signalling: a conserved mechanism regulating cell polarity and directed motility. *Nat Rev Genet.* **8**, 126–138.

3. Simons, M., and Mlodzik, M. (2008) Planar cell polarity signaling: from fly development to human disease. *Annu Rev Genet.* **42**, 517–540.

4. Tao, H., Suzuki, M., Kiyonari, H., Abe, T., Sasaoka, T., and Ueno, N. (2009) Mouse prickle1, the homolog of a PCP gene, is essential for epiblast apical-basal polarity. *Proc Natl Acad Sci USA* **106**, 14426–14431.

5. Fischer, E., and Pontoglio, M. (2009) Planar cell polarity and cilia. *Semin Cell Dev Biol.* **20**, 998–1005.

6. Andrew, D. J., and Ewald, A. J. (2009) Morphogenesis of epithelial tubes: Insights into tube formation, elongation, and elaboration. *Dev Biol.* **341**, 34–55.

7. Nishita, M., Enomoto, M., Yamagata, K., and Minami, Y. (2010) Cell/tissue-tropic functions of Wnt5a signaling in normal and cancer cells. *Trends Cell Biol.* **20**(6), 346–354.

8. Matsuyama, M., Aizawa, S., and Shimono, A. (2009) Sfrp controls apicobasal polarity and oriented cell division in developing gut epithelium. *PLoS Genet.* **5**, e1000427.

Chapter 19

Assessing PCP in the Cochlea of Mammalian Ciliopathy Models

Daniel J. Jagger and Andrew Forge

Abstract

The increased availability of mouse models of human genetic ciliary diseases has led to advances in our understanding of the diverse cellular roles played by cilia. The family of so-called "ciliopathies" includes Alström Syndrome, Bardet–Biedl Syndrome, Primary Ciliary Dyskinesia, and Polycystic Kidney Disease, among many others. In mouse models of Alström Syndrome and Bardet–Biedl Syndrome, we have shown developmental defects in the mechano-sensory stereociliary bundles on the apical surfaces of "hair" cells in the cochlea, the mammalian hearing organ. Stereocilia are specialized actin-based microvilli, whose characteristic patterning is thought to be dependent on the hair cell's primary cilium ("kinocilium"). Ciliopathy-associated proteins are localized to the ciliary axoneme and/or the ciliary basal body, or to the bundle itself. Ciliopathy-associated genes functionally interact with genes of the noncanonical Wnt pathway, and so implicate PCP in the control of hair cell development.

Key words: Alström Syndrome, Bardet–Biedl Syndrome, Cilium, Deafness, Hair cells, Kinocilium, Organ of Corti, Stereocilia

1. Introduction

The mammalian cochlea is responsible for the transduction of complex sound stimuli into meaningful neural code. Central to this process are mechano-sensory "hair" cells located in the auditory neuro-epithelium, the organ of Corti (1). Hair cells are held in a regular mosaic pattern along with surrounding epithelial "supporting" cells (2). Hair cells are characterized by the presence of an actin-based stereociliary bundle on their apical surface. During development the organization of the stereociliary bundle is likely determined by the positioning of a primary cilium (3). Positive mechanical deflection of the bundle results in hair cell depolarization and the release of neurotransmitter (inner hair cells) or voltage-dependent

Kursad Turksen (ed.), *Planar Cell Polarity: Methods and Protocols*, Methods in Molecular Biology, vol. 839,
DOI 10.1007/978-1-61779-510-7_19, © Springer Science+Business Media, LLC 2012

somatic motility resulting in sound amplification (outer hair cells). Normal hearing relies on a regular and stereotyped arrangement of the bundles, presenting the hair cell in the best orientation to sense mechanical stimuli. This patterning appears to be under the control of a large number of genes during development, and their mutation often results in clinically recognized forms of deafness (4).

Mouse models of the phenotypically similar human ciliopathies Alström Syndrome and Bardet–Biedl Syndrome are hearing-impaired (5, 6), and analysis of their hair cells reveals a range of peculiarities in the arrangement of their stereociliary bundles, and the positioning of the associated primary cilium (known as the kinocilium) (5, 7, 8). Thus, the cochlea provides a measurable "read-out" of the effects of gene deletion. Confocal immunofluorescence localizes disease-associated proteins to the ciliary axoneme and/or basal body in hair cells and their supporting cells (7, 8). Certain proteins are localized more widely along microtubules, which are uniquely concentrated within supporting cells. This association is shown to be reversible following the de-stabilization of microtubules by cold incubation of live cochlear slices. The identification of functional interactions between ciliopathy and PCP genes (5, 9) points toward additionally complex control mechanisms that determine the development of the organ of Corti.

2. Materials

2.1. Scanning Electron Microscopy

1. Cacodylate buffer: prepared as a stock solution 0.2 M sodium cacodylate (Sigma, Poole, UK) with 6 mM $CaCl_2$, pH 7.3. Store at 4°C. Buffer diluted 1:1 for working concentration of 0.1 M cacodylate, 3 mM $CaCl_2$.

2. Primary fixative: 2.5% glutaraldehyde in 0.1 M cacodylate buffer. Prepared from sealed vials of 25% glutaraldehyde in water (Agar Scientific, Stansted, UK). Dilute 1:5:4, 25% glutaraldehyde:0.2 M cacodylate:water. Store at 4°C.

3. Osmium tetroxide postfixative. Use OsO_4 inside a fume hood (see Note 1). A 1 g vial of solid OsO_4 (Agar Scientific, Stansted, UK), is opened inside a fume hood and added to 50 ml of distilled/de-ionized water to provide a 2% solution. The OsO_4 takes some time to dissolve. Best left overnight before first use. Can be stored at room temperature in dark-glass bottle to reduce exposure to light. For use mix equal volumes of 2% OsO_4 and 0.2 M cacodylate to give a 1% solution in 0.1 M cacodylate buffer as postfixative; dilute 1:1 with water during processing with thiocarbohydrazide.

4. Decalcification buffer: 4% EDTA (Sigma) in 0.1 M cacodylate buffer.

5. Thiocarbohydrazide (TCH) (Sigma): fresh, filtered saturated solution in water. 0.5 g per 100 ml provides a saturated solution. Filter before use. Use a fresh solution each time.

6. Critical point dry device: from liquid CO_2.

7. Specimen support stubs for SEM (Agar Scientific). Different instrument models require different types of specimen support stub.

8. Silver Paint (provides a conductive adhesive for attaching specimens to support stub): Electrolube Silver Conductive paint (RS Components); Electrodag fast drying conductive paint (Ted Pella Inc., Redding CA).

9. Sputter coater: Gold palladium or platinum foil (Agar Scientific).

2.2. Confocal Immunofluorescence for Primary Cilia and ALMS1

1. Phosphate buffer solution (PBS): prepared as a 1× solution with 137.0 mM NaCl, 2.7 mM KCl, 8.0 mM Na_2HPO_4, 1.5 mM KH_2PO_4, pH 7.3.

2. Paraformaldehyde (PFA) fixative (Fisher Scientific, Loughborough, UK): Prepare a 4% (w/v) solution in PBS, warming carefully on a stirring hotplate in a fume hood (see Note 2). Solution is cooled to room temperature before use, and can be aliquoted and stored at –20°C for up to 1 month.

3. Permeabilization/blocking buffer: 0.1% (v/v) Triton X-100 (Sigma) and 10% (v/v) normal goat serum (Sigma) in PBS. Keep refrigerated for 2–3 days.

4. Antibody dilution buffer: 0.1% (v/v) Triton X-100 and 100 mM L-lysine in PBS. Keep refrigerated for up to 1 month.

5. Primary antibodies: Mouse monoclonal anti-acetylated tubulin antiserum (Sigma, T6793); rabbit polyclonal anti-ALMS1 antiserum (10).

6. Secondary antibodies: Goat anti-mouse IgG conjugated to Alexa488 (Invitrogen); goat anti-rabbit IgG conjugated to Alexa555 (Invitrogen).

7. PTFE multispot glass slides (CA Hendley Essex Ltd, Loughton, UK), 22 × 50 mm microscope cover slips, nail varnish.

8. Anti-fade mountant: (Vectashield Mounting Medium with DAPI, Vector Labs, Burlingame CA).

2.3. Live Cochlear Slices for Cold-Stability of Microtubules

1. Artificial perilymph: 150 mM NaCl, 4 mM KCl, 2 mM $MgCl_2$, 1.5 mM $CaCl_2$, 10 mM Hepes, 5 mM D-glucose; adjust pH to 7.3 using 10 M NaOH. Make fresh every day.

2. Double-edged disposable razor blades (Boots Ltd, Nottingham, UK) (see Note 3).

3. Cyanoacrylate adhesive (Roti coll 1, Carl Roth GmbH, Karlsruhe, Germany) (see Note 4).

4. Paraformaldehyde (PFA) fixative: As above.

5. PBS: As above.

6. Permeabilization/blocking buffer: As above

7. Antibody dilution buffer: As above.

8. Primary antibody: Mouse monoclonal anti-acetylated tubulin antiserum (as above).

9. Secondary antibody: Goat anti-mouse IgG conjugated to Alexa488 (Invitrogen).

10. Cavity microslides (Agar Scientific, Stansted, UK), cover slips, nail varnish.

11. Anti-fade mountant, as above.

3. Methods

Genetic homozygous knockout (−/−) of ciliopathy-associated genes causes various effects on the development of the mouse cochlea. As these diseases are inherited in a recessive manner, wild-type (+/+) and heterozygous (+/−) littermates can be used as comparative controls. In practice, both ears are used for all experiments. The mouse or rat cochlea is largely cartilaginous at birth and bone calcification advances from around postnatal day 4 (P4), often necessitating a decalcification step before tissue dissection. Where possible rat cochleae are used for confocal immunofluorescence experiments, in order to circumvent "mouse-on-mouse" cross-reaction artifacts that can occur when using antibodies raised in mice on mouse tissues.

3.1. Scanning Electron Microscopy

1. In a 35-mm petri dish the bullae tympanica (the murine equivalents of human temporal bones) are dissected from the skull using fine tweezers. A bulla is opened and the cochlear bone is exposed. A small hole is made in the cochlear bone covering the apical turn region using a hypodermic needle, and the round window membrane punctured.

2. The cochlear tissues are fixed in situ by direct perfusion of glutaraldehyde solution. The fixative is injected gently through the openings at the basal and apical ends of the cochlea, then the entire opened auditory bulla immersed in glutaraldehyde solution for 2 h at room temperature on a mechanical rotator, or overnight at 4°C. Samples are washed at least three times for 10 min each wash in 0.1 M cacodylate buffer, then postfixed in buffered OsO_4 for 2 h at room temperature.

3. In mice older than P8, the cochlea is decalcified in EDTA solution for 48 h at 4°C (see Note 5).

4. The cochlea is dissected under cacodylate buffer and viewed with a stereo-dissecting microscope using incident illumination. In cochleae up to P8, the lateral wall tissues are relatively easily peeled away from the organ of Corti with fine forceps beginning at the basal coil. This exposes the entire organ of Corti spiral. The organ of Corti attached to the modiolus is divided into apical and basal segments by cutting through the modiolus. The tectorial membrane is lifted away using fine forceps and is peeled off. In older, decalcified cochleae, at the apical end the decalcified bone covering the lateral wall is gently cut away and the organ of Corti is exposed by removing segments of the lateral wall tissue with fine forceps and peeling away the tectorial membrane before cutting through the modiolus to release the apical half. For the basal coil, the decalcified bony wall and the lateral wall tissues are cut longitudinally, in the direction along cochlear spiral, at a level as close as possible to the surface of the organ of Corti. Again the tectorial membrane is peeled off the organ of Corti and the basal coil is then released by cutting the cochlea below the round window.

5. The dissected segments are processed through the repeated TCH-OsO$_4$ procedure (11). The samples are washed in distilled water, then incubated for 20 min at room temperature in a fresh, filtered, saturated solution of TCH. Following six washes in distilled water over about 30 min the samples are incubated in 1% aqueous OsO$_4$ for 1 h at room temperature followed by six washes in distilled water before repeating incubation in TCH and OsO$_4$ as before.

6. The samples are dehydrated through an alcohol series: 30%, 50%, 70%, 85%, and 95% ethanol for at least 10 min each and then 3×10 min incubation in pure 100% ethanol. The samples are then critical point dried from CO$_2$.

7. The dried organ of Corti segments are mounted on SEM support stubs using a fast drying silver paint. When the paint is fully dried they are lightly sputter-coated with gold palladium or platinum before examination. Examples of hair cell stereociliary bundles are shown in Fig. 1.

3.2. Confocal Immunofluorescence for Primary Cilia and ALMS1

1. In a 35-mm petri dish the bulla tympanica of a P4 rat is dissected from the skull using fine tweezers. The bulla is opened and the cartilaginous cochlear bone is exposed. A small hole is made in the cochlear bone covering the apical turn region using a hypodermic needle, and the round window membrane punctured.

2. The cochlear tissues are fixed in situ by direct perfusion of PFA solution through the round window. The cochlea is fixed in PFA solution for 30 min at room temperature. The cochlea is washed three times for 10 min each with PBS.

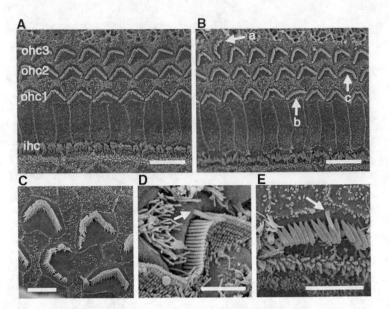

Fig. 1. Scanning electron microscopy reveals hair cell bundle peculiarities in the organ of Corti of a ciliopathy mouse model. (**a**) Surface view of the P11 *Bbs6*[+/+] mouse organ of Corti, revealing stereociliary bundles on inner hair cells (ihc) and three rows of outer hair cells (ohc1-3). The bundles of outer hair cells have a regular "V" or "W" shape. (**b**) In the organ of Corti of a *Bbs6*[-/-] littermate outer hair cells in all three rows show peculiarities in their bundle structure, including mis-orientation (*a*), flattening (*b*), and asymmetry (*c*). (**c**) Detail of further bundle peculiarities in a *Bbs6*[-/-] mouse. (**d**) In a P2 wild-type mouse the kinocilium (*arrow*) can be distinguished from the adjacent developing bundle on an outer hair cell. (**e**) In a P9 wild-type mouse the kinocilium (*arrow*) can be distinguished from the adjacent developing bundle on an inner hair cell. Scale bars, (**a**, **b**) 10 μm, (**c–e**) 5 μm.

3. Viewed under a binocular microscope and bathed in PBS, the organ of Corti is removed from the cochlea as a single piece using fine forceps. Reissner's membrane is removed and the tectorial membrane is lifted from the organ of Corti using fine forceps.

4. In a 96-well plate the organ of Corti is incubated in the permeabilization/blocking solution for 30 min at room temperature (see Note 6).

5. The permeabilization/blocking solution is removed and replaced with the anti-acetylated tubulin antibody (1:1,000) in antibody diluting buffer for 1 h at room temperature.

6. The primary antibody is removed and the sample washed three times for 10 min each with PBS. The sample is kept under aluminum foil for subsequent steps, to prevent excessive exposure to ambient light.

7. The secondary antibody is prepared at 1:400 in antibody dilution buffer and added to the sample for 30 min at room temperature.

8. The secondary antibody is removed and the sample is washed three times for 5 min each with PBS.

9. The sample is carefully transferred to a glass slide using fine forceps. The lumenal surface of the organ of Corti should be facing upward, for best visualization of stereociliary bundles and primary cilia. The sample can be subdivided into basal/middle/apical cochlear turns if necessary, and isolated in separate wells on the multispot glass slide. Small drops of mountant are added to each well, and a cover slip added (see Note 7). Nail varnish is used to seal the sample.

10. The sample is then viewed under epi-fluorescence microscopy, using the DAPI fluorescence (blue emission) to locate the characteristic rows of inner and outer hair cells. Under confocal microscopy the Alexa488 fluorescence (488 nm laser excitation, 505–530 nm band-pass emission) and Alexa555 fluorescence (543 nm laser excitation, 550 nm long-pass emission) are viewed and z-stack samples taken, and projected as necessary using computer software. Examples of signals for acetylated tubulin expression in organ of Corti primary cilia and cytoplasmic microtubules, and ALMS1 expression in ciliary basal bodies, are shown in Fig. 2.

3.3. Live Cochlear Slices for Cold-Stability of Microtubules

1. The bulla tympanica of a P4 rat is dissected from the skull using fine tweezers. Bathed in artificial perilymph, the bulla is opened and the calcifying cochlear bone is exposed. Care is taken to not break the cochlear bone or to puncture the membranes covering the round and oval windows.

2. The cochlea is carefully dried, and mounted on a vibratome block using a droplet (50 μl approx.) of cyanoacrylate adhesive.

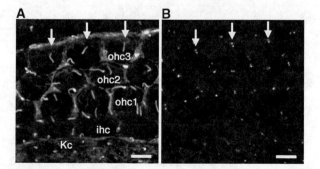

Fig. 2. Localization of ALMS1 in the neonatal rat organ of Corti. (a) Acetylated tubulin confocal immunofluorescence at the surface of the P2 rat organ of Corti reveals the cilia on the apical surface of hair cells and supporting cells. *Arrows* denote kinocilia on three outer hair cells in the third row. Abbreviations: *ihc* inner hair cell, *Kc* Kölliker's cells, *ohc1-3* outer hair cells. (b) ALMS1 confocal immunofluorescence in the tissue shown in (a), showing ubiquitous expression at the basal bodies of hair cells and supporting cells. *Arrows* denote ALMS1-labeled basal bodies in the outer hair cells highlighted in (a). Scale bars 5 μm.

The adhesive is set partially using a droplet of artificial perilymph, and the cochlea is positioned quickly to ensure that the round window is facing upward and that the modiolar axis of the cochlea will be parallel to the vibratome blade when the block is inserted. The block and mounted cochlea are submerged under artificial perilymph to set the adhesive completely (see Note 8).

3. The block is inserted into the vibratome (Vibratome 1000+, Technical Products International Inc., St Louis, MO) and a razor blade is fitted into the blade holder. Serial slices (200 μm approx.) are taken, with the blade advance speed set appropriately slowly and the blade displacement minimized such that slices are cut cleanly, without splintering of the cochlear bone (see Note 9), even when calcified slices are prepared (12). Slices are maintained in artificial perilymph at room temperature until required.

4. For cold-disruption of cytoplasmic microtubules, slices are maintained at 4°C for 5 h, control slices are maintained at room temperature. Slices are then fixed in PFA solution for 1 h at the test temperature and a further 1 h at room temperature.

5. Slices are incubated in the permeabilization/blocking solution for 30 min at room temperature.

6. The permeabilization/blocking solution is removed and replaced with the anti-acetylated tubulin antibody (1:1,000) in antibody diluting buffer for 1 h at room temperature.

7. The primary antibody is removed and the sample washed three times for 10 min each with PBS. The sample is kept under aluminum foil for subsequent steps to prevent excessive exposure to ambient light.

8. The secondary antibody is prepared at 1:400 in antibody dilution buffer and added to the sample for 30 min at room temperature.

9. The secondary antibody is removed and the sample washed three times for 5 min each with PBS.

10. Each slice is carefully transferred to individual cavity microslides using fine forceps, mountant applied and a cover slip added. Nail varnish is used to seal the sample.

11. Slices are viewed under epi-fluorescence microscopy, using the DAPI fluorescence (blue emission) to locate the characteristic rows of inner and outer hair cells. Under confocal microscopy the Alexa488 fluorescence (488 nm laser excitation, 505–530 nm band-pass emission) is viewed and z-stack samples taken, and projected as necessary using computer software (see Note 10). Examples of signals for acetylated tubulin expression in organ of Corti primary cilia and cytoplasmic microtubules are shown in Fig. 2.

4. Notes

1. Osmium tetroxide is extremely volatile and a hazard to human health.

2. Paraformaldehyde is a possible carcinogen and great care should be taken.

3. We have used several brands of stainless steel razor blades to prepare cochlear slices. We have not identified a particular brand that is noticeably superior to others. The blade can be moved along in the blade holder between cuts, in order to utilize a sharp region of the cutting edge. This is particularly helpful when cutting ossified tissue that can damage the cutting edge. The blade angle is set at 20°.

4. We have used several brands of cyanoacrylate adhesive, with varying amounts of success. We find that certain brands set too hard, too quickly. Others expand on being submerged in physiological saline. We have had considerable success using the brand listed, though performance seems to vary between batches.

5. Decalcification after osmium fixation is recommended although it can be performed after the initial glutaraldehyde fixation. If decalcification follows the OsO_4 postfixation step, there is somewhat better preservation of tissue and the subsequent dissection is easier to perform because the tissues are stained brown-black and therefore easier to see.

6. The organ of Corti can be subdivided into basal, mid, and apical turns if preferred. This allows comparison of immunofluorescence in different frequency-coding locations. However, if the tissue is to be incubated in the same antibody solutions, we find that it is easier to handle the tissue in a single piece, and subdivide it prior to mounting.

7. Organ of Corti may be flattened somewhat when the cover slip is lowered into the mountant. To minimize the distortion of stereocilia and kinocilia, we have found it is possible to increase the effective depth of the PTFE wells by painting nail varnish "walls" around the wells before adding the cover slip.

8. As the cyanoacrylate sets it is possible to shape it with blunt forceps, in order to avoid large masses of glue that can blunt the blade before it can cut the cochlea.

9. Internal potentiometers on the Vibratome 1000+ can be adjusted to set the ranges of desired blade advance speed and blade displacement (vibration). We recommend that users take care to adjust the vibratome to a low range of blade advancement to achieve a suitably slow working speed.

10. To image thicker sections (~200 μm) we use a 63× water-immersion objective, which has a longer working distance than oil-immersion objectives of comparable magnification. Alternatively, we use 20× air-immersion objectives, though these have a relatively poor numerical aperture (N.A.) and are not ideal to image single cells.

Acknowledgments

The authors would like to thank Phil Beales (UCL) for *Bbs* mice, and David Wilson (University of Southampton) for the ALMS1 antibody. The Biotechnology and Biological Sciences Research Council and Deafness Research UK supported this work. DJ is a Royal Society University Research Fellow.

References

1. Forge, A., Wright, T.: The molecular architecture of the inner ear. *Br Med Bull* **63**, 5–24 (2002).

2. Gale, J.E., Jagger, D.J.: Cochlear supporting cells. In: Fuchs, P.A. (ed.) The Oxford Handbook of Auditory Science: The Ear. pp. 307–327. Oxford University Press, Oxford (2010).

3. Denman-Johnson, K., Forge, A.: Establishment of hair bundle polarity and orientation in the developing vestibular system of the mouse. *J Neurocytol* **28**(10–11), 821–835 (1999).

4. Nayak, G.D., Ratnayaka, H.S., Goodyear, R.J., Richardson, G.P.: Development of the hair bundle and mechanotransduction. *Int J Dev Biol* **51**(6–7), 597–608 (2007). doi:072392gn [pii] 10.1387/ijdb.072392gn.

5. Ross, A.J., May-Simera, H., Eichers, E.R., Kai, M., Hill, J., Jagger, D.J., Leitch, C.C., Chapple, J.P., Munro, P.M., Fisher, S., Tan, P.L., Phillips, H.M., Leroux, M.R., Henderson, D.J., Murdoch, J.N., Copp, A.J., Eliot, M.M., Lupski, J.R., Kemp, D.T., Dollfus, H., Tada, M., Katsanis, N., Forge, A., Beales, P.L.: Disruption of Bardet-Biedl syndrome ciliary proteins perturbs planar cell polarity in vertebrates. *Nat Genet* **37**(10), 1135–1140 (2005).

6. Collin, G.B., Cyr, E., Bronson, R., Marshall, J.D., Gifford, E.J., Hicks, W., Murray, S.A., Zheng, Q.Y., Smith, R.S., Nishina, P.M., Naggert, J.K.: Alms1-disrupted mice recapitulate human Alstrom syndrome. *Hum Mol Genet* **14**(16), 2323–2333 (2005).

7. May-Simera, H.L., Ross, A., Rix, S., Forge, A., Beales, P.L., Jagger, D.J.: Patterns of expression of Bardet-Biedl syndrome proteins in the mammalian cochlea suggest noncentrosomal functions. *J Comp Neurol* **514**(2), 174–188 (2009).

8. Jagger, D., Collin, G., Kelly, J., Towers, E., Nevill, G., Longo-Guess, C., Benson, J., Halsey, K., Dolan, D., Marshall, J., Naggert, J., Forge, A.: Alstrom Syndrome protein ALMS1 localizes to basal bodies of cochlear hair cells and regulates cilium-dependent planar cell polarity. *Hum Mol Genet* **20**(3), 466–481 (2011). doi:ddq493 [pii] 10.1093/hmg/ddq493.

9. Jones, C., Chen, P.: Primary cilia in planar cell polarity regulation of the inner ear. *Curr Top Dev Biol* **85**, 197–224 (2008).

10. Hearn, T., Spalluto, C., Phillips, V.J., Renforth, G.L., Copin, N., Hanley, N.A., Wilson, D.I.: Subcellular localization of ALMS1 supports involvement of centrosome and basal body dysfunction in the pathogenesis of obesity, insulin resistance, and type 2 diabetes. *Diabetes* **54**(5), 1581–1587 (2005).

11. Davies, S., Forge, A.: Preparation of the mammalian organ of Corti for scanning electron microscopy. *J Microsc* **147**(Pt 1), 89–101 (1987).

12. Jagger, D.J., Forge, A.: Compartmentalized and signal-selective gap junctional coupling in the hearing cochlea. *J Neurosci* **26**(4), 1260–1268 (2006).

Chapter 20

Morphometric Analysis of Centrosome Position in Tissues

Hester Happé, Emile de Heer, and Dorien J.M. Peters

Abstract

Planar cell polarity (PCP) is the polarization of cells within the plane of an epithelial cell layer. PCP is important in many tissues in different processes. In the kidney, it is hypothesized to be important in acquiring and maintaining correct tubular diameter. Aberrant PCP has been shown to be involved in polycystic kidney disease. Therefore, research in this field requires a method to study PCP. As PCP and outward-in signaling via the cilia are interconnected, the position of the centrosome, the base of the cilium can be used as a read-out system for PCP. Here, we provide a method in which the position of the centrosome is measured as read-out for PCP.

Key words: Planar cell polarity, Centrosome position, Cilium, Morphometric analysis, Immunofluorescence

1. Introduction

Monolayers of epithelial cells exhibit, in addition to apical-basal polarity, also planar cell polarity. Planar cell polarity (PCP) is the coordinated polarization of cells in the plane of a tissue, thereby referring to asymmetries within the plane of the epithelial cell layer. PCP is required for many cellular processes. For example, it is thought to play an important role in acquiring and maintaining correct tubular morphology during renal development as well as during repair after epithelial injury. Aberrant PCP can cause cyst formation and is altered in polycystic kidney disease (1–5).

PCP and (primary) cilia exert a mutual interaction in maintaining cellular functions.

The PCP pathway plays a role in regulating the position and orientation of cilia. However, when cilia orient to flow direction, orientation of cilia can also regulate PCP. Cilia consist of a microtubule-based axoneme covered by a specialized plasma membrane that extends from the cell surface into the extracellular space.

Kursad Turksen (ed.), *Planar Cell Polarity: Methods and Protocols*, Methods in Molecular Biology, vol. 839,
DOI 10.1007/978-1-61779-510-7_20, © Springer Science+Business Media, LLC 2012

The basal body/centrosome forms the base of the cilium. The position of the centrosome indicates the position of the cilium and deviation of the normal position in the apical membrane may indicate altered PCP.

Another frequently used read-out, to study aberrant PCP is a measurement of the orientation of the nuclear spindle, with the aim to measure alterations in oriented cell division. This analysis can only be done in proliferative tissue when many dividing cells are present, to obtain a sufficient number of measurable cells in anaphase.

Measuring centrosome position in tissues to indicate altered PCP was first described by Jonassen et al. (5). We modified the protocol to study centrosome localization in a model for polycystic kidney disease. We showed that at precystic stages renal tubular epithelial cells display a significantly different distribution of centrosome position as compared to control mice (4). In the majority of renal epithelial cells of control mice, centrosomes are positioned close to the center of the apical membrane. In mutants, however, a significantly different pattern was observed.

The following protocol describes the morphometric analysis of centrosome position in renal tubular epithelial cells in formalin-fixed paraffin-embedded renal tissue sections that is applicable to nonproliferative tissue.

2. Materials

2.1. Immuno-fluorescence

1. Superfrost + microscope slides.

2. Xylene.

3. Ethanol, anhydrous denatured, histological grade (100% and 95% and 70%).

4. Deionized water (dH_2O).

5. Buffer for heat-mediated antigen retrieval 10/1 mM TRIS/EDTA, pH 9.0.

   To prepare 1 l add 1.21 g Trizma® base ($C_4H_{11}NO_3$) and 0.372 g EDTA ($C_{10}H_{14}N_2O_8Na_2 \cdot 2H_2O$) to 950 ml dH_2O. Adjust pH to 9.0, then adjust final volume to 1,000 ml with dH_2O. Store at 4°C.

6. Phosphate-buffered saline (PBS).

   To prepare 1 l 10× stock: add 80 g sodium chloride (NaCl), 2 g potassium chloride (KCl), 14.4 g sodium phosphate, dibasic (Na_2HPO_4), and 2.4 g potassium phosphate, monobasic (KH_2PO_4) to 1 l dH_2O. Adjust pH to 7.4 (adjust to pH 7.4 with HCl if necessary). Prepare working solution by dilution of one part with nine parts of water.

7. Antibody dilution buffer: 1% (w/v) bovine serum albumin (BSA) fraction V in PBS.

8. Normal goat serum (NGS).

9. Primary antibody: rabbit polyclonal anti-collagen IV IgG (Abcam).

10. Primary antibody: mouse monoclonal anti-g-tubulin IgG1 (clone GTU-88, Sigma-Aldrich).

11. Secondary antibody: goat-anti-rabbit IgG Alexa546 (Invitrogen).

12. Secondary antibody: goat-anti-mouse IgG1 Alexa 488 (Invitrogen).

13. Vectashield containing DAPI.

2.2. Image Acquisition and Morphometric Analysis

1. Immersion oil.

2. Leica DM500B microscope with a Leica DM DFC 350 FX camera using Leica Application Suite 1.8.0.

3. Image J software (public domain software, NIH, USA; http://rsb.info.nih.gov/ij/).

3. Methods

To measure the position of the centrosome, sections will be stained with γ-tubulin to indicate the centrosome, collagen IV to indicate the basement membrane, and DAPI for the nuclei. The position of the centrosome is measured by the angle defined by the line through the center of the nucleus perpendicular to the basement membrane and a line through the centrosome toward the intersection of the first line at the basement membrane (see Fig. 1).

Regarding PCP, localization of the centrosome can be affected in two directions. In renal tubules, centrosome position can be changed parallel to the longitudinal axis of the tubule or perpendicular to it. Therefore, only true longitudinal or transversal tubular cross sections should be measured. Otherwise, measurements will be the result of the vector of both directions. As deviations may occur only parallel to the longitudinal axis of the tubule or only perpendicular to it, one should chose to measure deviations only in one direction, or to keep measurements for the two directions separate. Pooling data of measurements in two directions could result in a loss of information. When deviations in the longitudinal direction are considered, it is important to keep track of the orientation of the tubules in such a way that for all pictures it is clear what is the proximal and what is the distal side.

Here, we only considered deviations in centrosome position perpendicular to the tubular axis.

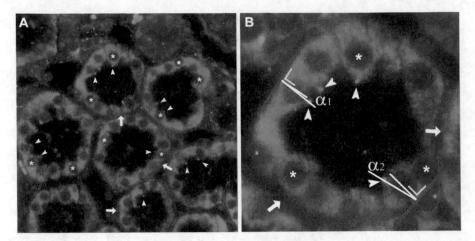

Fig. 1. Measurement of centrosome position in renal tubular epithelial cells. Kidney cortex section stained for γ-tubulin for centrosomes (*green*), type IV collagen for basement membrane (*red*), and DAPI-stained nuclei (*blue*). Examples of centrosomes, nuclei, and basal membranes are indicated by *arrowheads*, *asterisks*, and *arrows*, respectively. (**a**) Image of renal tubules with adjusted color balance. (**b**) The position of the centrosome was determined by measuring the angle between a line through the center of the nucleus perpendicular to the basement membrane and a line through the centrosome toward this point at the basement membrane. Autofluorescence is detected in the same channel as γ-tubulin. Centrosome angles are indicated by α_1 and α_2.

3.1. Immunofluorescence

Do not allow slides to dry at any time after step 1 during this procedure. Drying of the sections will result in high background staining. All incubation steps should be performed in incubation boxes with wetted (filter) paper on the bottom avoiding evaporation of the incubation liquid and drying of the sections. Ideally, glass slides should not directly lie on the wet paper.

1. Collect formalin-fixed paraffin-embedded kidney sections (4 μm) on Superfrost+ glass slides and dry o/n at 37°C.

2. Deparaffination/rehydration

   (a) Incubate sections in three washes of xylene for 5 min each.

   (b) Incubate sections in two washes of 100% ethanol for 5 min each.

   (c) Incubate sections in two washes of 95% ethanol for 5 min each.

   (d) Incubate sections in two washes of 70% ethanol for 5 min each.

   (e) Wash sections twice in dH$_2$O for 5 min each.

3. Carefully boil slides for 10 min in 10/1 mM TRIS/EDTA using a microwave oven, and allow cooling down for 20 min (see Notes 1 and 2).

4. Put circles around sections using DAKO-Pen to make sure the antibody solution stays on the section.

5. Preincubate with 50 µl 5% (v/v) NGS in antibody dilution buffer for 30 min at room temperature (see Note 3).

6. Discard blocking solution by tapping the glass slide on a tissue and wipe off the rest of the liquid (of course do not touch the section itself).

7. Prepare a primary antibody mixture containing rabbit poly-clonal anti-collagen IV IgG1 diluted 1:500 and mouse mono-clonal anti-γ-tubulin IgG1 diluted 1:750 in antibody dilution buffer.

8. Incubate with 50 µl of the primary antibody mixture o/n at 4°C with. Discard primary antibody solution.

9. Rinse slides three times for 5 min with PBS.

10. Remove as much PBS from the glass slides as possible.

11. Prepare secondary antibody solution: antibody dilution buffer containing goat-anti-rabbit IgG Alexa546 (1:200) and goat-anti-mouse IgG1 Alexa 488 (1:200) (keep in the dark).

12. Incubate sections with 50 µl of the secondary antibody solution for 1 h at room temperature. Keep the slides from now on away from light as much as possible to avoid loss of signal from the Alexa Dyes.

13. Wash 3× 5 min with PBS.

14. Remove as much of the PBS as possible with filter paper without touching the section and air dry the section for a few minutes.

15. Mount slides with Vectashield containing DAPI.

3.2. Image Acquisition and Morphometric Analysis

1. The slides are viewed using fluorescent microscopy. Excitation at 556 nm induces fluorescence of the Alexa 546 dye at 573 nm (orange emission) for the collagen IV, while excitation at 495 nm induces fluorescence of the Alexa 488 dye at 519 nm (green emission) for γ-tubulin, marker for centrosomes. Excitation at 345 nm induces DAPI fluorescence at 458 (blue emission) for nuclei.

2. Make Z-stacks, of 25–30 Z-images taken 0.24 mm apart, of the area in the section you are interested in using 63× oil-immersion lens with immersion oil (see Notes 4 and 5).

3. Take 25–35 different pictures of the area of interest, in order to measure appropriate numbers of centrosome angles.

4. Convert the Z-stack to a maximum projection.

5. Import images into Image J and adjust the color balance in order to make the centrosomes clearly visible (Fig. 1a shows an example of a maximum projection with adjusted color balance).

6. Enlarge the picture if needed.

7. Select the tool for measuring angles in Image J.

8. Measuring the angle between a line through the center of the nucleus perpendicular to the basement membrane and a line through the centrosome towards this point at the basement membrane. Collagen IV staining was used to indicate the basement membrane (see Fig. 1b).

   Crl + M will put your measurement in a results screen. You can save this results file and open it in Excel.

   In Image J you have to "drag and drop" these two lines, by clicking at three points in the picture immediately after one another, you cannot adjust one of the two lines afterward. To reduce the effect of natural occurring variance and measuring errors, many angles have to be measured, 200–300 angles per group (three kidneys per group). It is also wise to do a blind analysis.

   Only measure centrosomes which you are sure that they belong to the same cell for which you are drawing a line through its nucleus (see Note 6).

9. When analyzing measurements and performing statistics it is most clear and informative to look at the distribution pattern of the centrosome angles. Differences in distribution can be tested with the Kolomogorov–Smirnov Z-test.

4. Notes

1. The antigen retrieval method is dependent on the specific antibodies, double staining is only possible when both antibodies give sufficient result with the same antigen retrieval method. Only one type of AR can be applied. We have found these antibodies to both give good signal using TRIS/EDTA antigen retrieval.

2. First let buffer cook in microwave at high power, put slides in a rack in the buffer, and maintain subboiling temperature in microwave oven (at lower power).

3. Using a volume of 50 µl is usually enough to cover the section, use more when having large sections.

4. In order to get as much centrosomes in focus in one picture as possible, it is preferable to use a microscope that is able to make Z-stacks, followed by a maximum projection of this Z-stack. It is possible to use a microscope without this function, however, the amount of clearly visible centrosomes will be lower as they are small and not all in the same plane of focus.

5. We acquired images of areas within the inner cortex of the kidney, containing transversally sectioned proximal tubules. You can use the collagen IV staining as well as auto-fluorescence as a guide for the location in the kidney.

6. If you are measuring longitudinal cross sections of tubules, make sure you know what is proximal and what is distal and to assign a plus or minus sign to the direction of the angle. Image J will not do this, it measures absolute angles.

Acknowledgments

We thank A. van der Wal and F. Prins for technical assistance.

References

1. Fischer E, Legue E, Doyen A, Nato F, Nicolas JF, Torres V, Yaniv M, Pontoglio M (2006) Defective planar cell polarity in polycystic kidney disease. *Nat Genet* **38**:21–23.

2. Patel V, Li L, Cobo-Stark P, Shao X, Somlo S, Lin F, Igarashi P (2008) Acute kidney injury and aberrant planar cell polarity induce cyst formation in mice lacking renal cilia. *Hum Mol Genet* **17**:1578–1590.

3. Saburi S, Hester I, Fischer E, Pontoglio M, Eremina V, Gessler M, Quaggin SE, Harrison R, Mount R, McNeill H (2008) Loss of Fat4 disrupts PCP signaling and oriented cell division and leads to cystic kidney disease. *Nat Genet* **40**:1010–1015.

4. Happe H, Leonhard WN, Van der Wal A, Van de Water B, Lantinga van Leeuwen IS, Breuning MH, De Heer E, Peters DJ (2009) Toxic tubular injury in kidneys from Pkd1-deletion mice accelerates cystogenesis accompanied by dysregulated planar cell polarity and canonical Wnt signaling pathways. *Hum Mol Genet* **18**:2532–2542.

5. Jonassen JA, San Agustin J, Follit JA, Pazour GJ (2008) Deletion of IFT20 in the mouse kidney causes misorientation of the mitotic spindle and cystic kidney disease. *J Cell Biol* **183**:377–384.

INDEX

Kursad Turksen (ed.), *Planar Cell Polarity: Methods and Protocols*, Methods in Molecular Biology, vol. 839,
DOI 10.1007/978-1-61779-510-7, © Springer Science+Business Media, LLC 2012